SOLOMON ACADEMY   Distribution or replication of part of this page is prohibite

Solomon Academy

PRACTICE TESTS FOR

I A A T

IOWA ALGEBRA APTITUDE TEST

by Yeon Rhee

Key Summaries for the IOWA Algebra Aptitude Test

7 Full-Length Practice Tests

Detailed Solutions for All Questions

www.solomonacademy.net

SOLOMON ACADEMY
Distribution or replication of any part of this page is prohibited.
LEGAL NOTICE

Legal Notice

IOWA Algebra Aptitude Test was not involved in the production of this publication nor endorses this book.

Copyright © 2014 by Solomon Academy
Published by: Solomon Academy
First Edition
ISBN-13: 978-1500258153
ISBN-10: 1500258156

All rights reserved. This publication or any portion thereof may not be copied, replicated, distributed, or transmitted in any form or by any means whether electronically or mechanically whatsoever. It is illegal to produce derivative works from this publication, in whole or in part, without the prior written permission of the publisher and author.

Acknowledgements

I wish to acknowledge my deepest appreciation to my wife, Sookyung, who has continuously given me wholehearted support, encouragement, and love. Without you, I could not have completed this book.

Thank you to my sons, Joshua and Jason, who have given me big smiles and inspiration. I love you all.

Thank you to my wonderful editor, Daniel Kwon, who has given me great advice and invaluable help.

About This Book

This book is designed towards mastering the Iowa Algebra Aptitude Test (IAAT), a placement test which allows students to demonstrate their readiness and ability to succeed in Algebra 1. The book contains a review of essential theorems specific to the IAAT: Pre-Algebraic Number Skills and Concepts, Mathematical Data Interpretation and Analysis, Representing Relationships, and Symbols. There are 7 full-length math tests with detailed solutions and explanations for all questions.

Be sure to time yourself during the mathematic test with the appropriate time limit of 40 minutes. After completing any lessons or tests, immediately use the answer key and detailed solution to check your answers. Review all answers. Take the time to carefully read the explanations of problems you got incorrect. If you find yourself continually missing the same type of questions, look back at the topic summaries and review the theorems and examples in the lesson. Set a goal of improvement for each practice test.

About Author

Yeon Rhee obtained a Masters of Arts Degree in Statistics at Columbia University, NY. He served as the Mathematical Statistician at the Bureau of Labor Statistics, DC. He is the Head Academic Director at Solomon Academy due to his devotion to the community coupled with his passion for teaching. His mission is to help students of all confidence level excel in academia to build a strong foundation in character, knowledge, and wisdom.

About IOWA Algebra Aptitude Test

The Iowa Algebra Aptitude Test (IAAT) is a placement test which allows students to demonstrate their readiness and ability to succeed in Algebra I. Students who pass with at least 91% percentile in the IAAT test will be granted the opportunity to take Algebra I in 7th grade. Notice that 91% percentile does not mean 91% correct; the score needed to qualify for Algebra I differs each year.

The Iowa Algebra Aptitude Test (IAAT) is a timed test. Although the administered time is approximately 70 minutes, the actual test-taking time is 40 minutes. The IAAT consists of 60 questions divided into four subtests; each subtest will be allotted 10 minutes for completion.

Pre-Algebraic Number Skills and Concepts
This subtests measures the student's capability to solve mathematics problems, understand Pre-Algebraic concepts and basic computational skills: Whole Numbers, Fractions and Mixed Numbers, Decimals and Percents, Estimation, and Ratios.

Mathematical Data Interpretation and Analysis
This subtests measures the student's capability to learn new information presented in text or graphs.

Representing Relationships
The student must demonstrate the ability to find the formulas and rules of a given numerical relationship whether depicted verbally or in table.

Symbols
Students must understand important symbols of algebra and how to apply them. This subtest includes solving algebraic expression and applying symbolic representation.

Contents

IAAT Topic Summaries	10
Practice Test 1 Section 1	25
Practice Test 1 Section 2	28
Practice Test 1 Section 3	32
Practice Test 1 Section 4	35
Practice Test 1 Section 1 Answers and Solutions	38
Practice Test 1 Section 2 Answers and Solutions	41
Practice Test 1 Section 3 Answers and Solutions	44
Practice Test 1 Section 4 Answers and Solutions	48
Practice Test 2 Section 1	52
Practice Test 2 Section 2	55
Practice Test 2 Section 3	58
Practice Test 2 Section 4	61
Practice Test 2 Section 1 Answers and Solutions	64
Practice Test 2 Section 2 Answers and Solutions	67
Practice Test 2 Section 3 Answers and Solutions	70
Practice Test 2 Section 4 Answers and Solutions	74
Practice Test 3 Section 1	77
Practice Test 3 Section 2	80
Practice Test 3 Section 3	84
Practice Test 3 Section 4	87

SOLOMON ACADEMY

TABLE of CONTENTS

Distribution or replication of any part of this page is prohibited.

Practice Test 3 Section 1 Answers and Solutions	90
Practice Test 3 Section 2 Answers and Solutions	93
Practice Test 3 Section 3 Answers and Solutions	96
Practice Test 3 Section 4 Answers and Solutions	100
Practice Test 4 Section 1	104
Practice Test 4 Section 2	107
Practice Test 4 Section 3	111
Practice Test 4 Section 4	114
Practice Test 4 Section 1 Answers and Solutions	117
Practice Test 4 Section 2 Answers and Solutions	120
Practice Test 4 Section 3 Answers and Solutions	123
Practice Test 4 Section 4 Answers and Solutions	126
Practice Test 5 Section 1	129
Practice Test 5 Section 2	132
Practice Test 5 Section 3	136
Practice Test 5 Section 4	139
Practice Test 5 Section 1 Answers and Solutions	142
Practice Test 5 Section 2 Answers and Solutions	145
Practice Test 5 Section 3 Answers and Solutions	147
Practice Test 5 Section 4 Answers and Solutions	151
Practice Test 6 Section 1	155
Practice Test 6 Section 2	158
Practice Test 6 Section 3	162
Practice Test 6 Section 4	165
Practice Test 6 Section 1 Answers and Solutions	168
Practice Test 6 Section 2 Answers and Solutions	171
Practice Test 6 Section 3 Answers and Solutions	174
Practice Test 6 Section 4 Answers and Solutions	178

SOLOMON ACADEMY

Distribution or replication of any part of this page is prohibited.

TABLE of CONTENTS

Practice Test 7 Section 1	181
Practice Test 7 Section 2	184
Practice Test 7 Section 3	188
Practice Test 7 Section 4	192
Practice Test 7 Section 1 Answers and Solutions	195
Practice Test 7 Section 2 Answers and Solutions	199
Practice Test 7 Section 3 Answers and Solutions	202
Practice Test 7 Section 4 Answers and Solutions	205

IAAT TOPIC SUMMARIES

Factors

Factors are the numbers that you multiply to get another number. For instance, $10 = 1 \times 10$ and $10 = 2 \times 5$. Thus, the factors of 10 are 1, 2, 5, and 10.

Prime and Composite Numbers

A **prime number** is a whole number that has only two factors: 1 and itself. The prime numbers less than 60 are 2, 3, 5, 7, 11, 13, 17, 19, 23, 29, 31, 37, 41, 43, 47, 53, and 59. It is worth noting that 2 is the first, smallest, and only even prime number among the prime numbers. A **composite number** is a number that has more than two factors. For instance, 4 has three factors and is composite: 1, 2, and 4. 0 and 1 are neither prime nor composite.

Prime Factorization

A **prime factorization** of a number is the process of writing the number as the product of all its prime factors. Since 12 can be written as $12 = 2 \times 2 \times 3$, the prime factorization of 12 is $2^2 \times 3$.

Remainder

A **remainder**, r, is the amount left over when a number is divided by a divisor. For instance, when 7 is divided by 2, the quotient is 3 and the remainder is 1. The remainder must always be less than the divisor but greater than or equal to 0.

The Order of Operations

The order of operations is used to simplify or evaluate numerical expressions. The order of operations is depicted by the acronym, PEMDAS: P stands for parenthesis, E stands for exponent, M stands for multiplication, D stands for division, A stands for addition, and S stands for subtraction.

The order of operations suggests to first perform any calculations inside parentheses. Afterwards, evaluate any exponents. Next, perform all multiplications and divisions working from left to right. Finally, do additions and subtractions from left to right. The example below shows how to evaluate numerical expressions using PEMDAS.

$$12(3-4)^2 \div 4 - 2 = 12(-1)^2 \div 4 - 2$$
$$= 12 \div 4 - 2$$
$$= 3 - 2$$
$$= 1$$

Adding and Subtracting Decimals

When adding and subtracting numbers containing decimal places, write down the numbers vertically and line up decimal points. Place zeros into the numbers so that all numbers have the same number of decimal places. Then, proceed to evaluate. For example, $4.05 + 3.1 + 2.055 = 9.205$

```
    4.05
    3.1
+   2.055
   ------
    9.205
```

Multiplying Decimals

When multiplying numbers containing decimal points, multiply ignoring the decimal points. Afterwards, place a decimal point into the product same as the total number of decimal places the two factors have together. For instance, let's multiply 5×0.4. First, convert 5×0.4 into $5 \times 4 = 20$ by removing all decimal points. Since 5 has no decimal places and 0.4 has one number after the decimal point, the product of 5×0.4 will have one decimal place. Thus, $5 \times 0.4 = 2$.

Dividing Decimals

To divide decimals, it is necessary to convert the divisor, or the number you are dividing by, into a whole number. The number of decimal places that the decimal point moves in the divisor must be mirrored by the dividend. Proceed to use long division to evaluate and add a decimal point in the same spot as the dividend. For example, $25 \div 1.6 = 250 \div 16$. After you have converted the divisor into a whole number, it is possible to use long division; $25 \div 1.6 = 250 \div 16 = 15\frac{5}{8}$.

Fractions

A fraction is a number that represents a part of a whole; all parts are equal to each other. The top number is the numerator and the bottom number is the denominator. There are three types of fractions: Proper Fractions, Improper Fractions, and Mixed Fractions. Mixed Fractions are commonly referred to as Mixed Numbers. A Proper Fraction is defined when the numerator is less than the denominator. An Improper Fraction has a numerator that is greater than (or equal to) the denominator. A Mixed Fraction, or Mixed Number, consists of a whole number and a proper fraction together.

$$\text{Examples of Proper Fractions are } \frac{2}{3}, \frac{5}{11}, \frac{24}{25}$$

$$\text{Examples of Improper Fractions are } \frac{3}{2}, \frac{11}{5}, \frac{25}{24}, \frac{3}{3}$$

$$\text{Examples of Proper Fractions are } 1\frac{1}{2}, 2\frac{1}{5}, 1\frac{1}{25}$$

Simplifying Fractions

To simplify a fraction, divide the top and bottom by the greatest common factor, or by the highest number that can divide into both numbers exactly. In order to simplify $\frac{30}{54}$, it is necessary to find the greatest common factor, or GCF, of 30 and 54. To find the greatest common factor, it is possible to use two different methods: listing out all the factors of each number and comparing or prime factorization. The prime factorization method is different from the least common multiple; therefore, be aware and do **NOT** get them mixed up.

Factors of 30: 1, 2, 3, 5, **6**, 10, 15, 30
Factors of 54: 1, 2, 3, **6**, 9, 18, 27, 54
Since 6 is the largest common factor of 30 and 54, 6 is the GCF.

The prime factorization of 30: $2 \times 3 \times 5$
The prime factorization of 54: $2 \times 3 \times 3 \times 3$
To find the GCF from the prime factorization, multiply the terms that they have in common. Although 54 has three 3's, 30 only has one 3 so both 30 and 54 only have one 3 in common. Since both 30 and 54 have one 2 and one 3 in common, the GCF is the product of all common prime factors: $2 \times 3 = 6$.

After obtaining the greatest common factor, divide both the numerator and demonitator of the fraction by the GCF to obtain the simplest form.

$$\frac{30}{54} = \frac{30 \div 6}{54 \div 6} = \frac{5}{9}$$

Adding and Subtracting Fractions

When adding and subtracting fractions, it is necessary to have a common denominator. If the fractions have a common denominator, proceed to add. Only the numerator increases and decreases in value as the denominator stays the same because the denominator represents how many parts represent a whole. For example, $\frac{1}{4} + \frac{2}{4} = \frac{3}{4}$.

When adding and subtracting fractions with unlike denominators, it is necessary to find the least common denominator. The least common denominator is the least common multiple of the denominators which can be found by listing out the multiples for each number or by a prime factorization method. For example, evaluate $\frac{3}{24} + \frac{5}{56}$. Since the denominators are 24 and 56, let's find the least common multiple of 24 and 56.

Multiples of 24: 24, 48, 72, 96, 120, 144, **168**, 192, $\cdots$
Multiples of 56: 56, 112, **168**, 224, $\cdots$
Since 168 is the smallest common multiple of 24 and 56, 168 is the LCM.

The prime factorization of 24: $2 \times 2 \times 2 \times 3 = 2^3 \times 3$
The prime factorization of 56: $2 \times 2 \times 2 \times 7 = 2^3 \times 7$
Since the prime factors of both numbers consist of 2, 3, and 7, multiply the greatest power of 2, 3, and 7 that appear in either prime factorization. Since both numbers have 2^3, it doesn't matter and 2^3 is the greater factorization of 2. 3 only appears in the prime factorization of 24 and, likewise, 7 only appears in the prime factorization of 56. Therefore, the LCM of 24 and 56 is $2^3 \times 3 \times 7 = 168$.

Since the least common multiple is 168, it is necessary to convert each fraction to have a denominator of 168 by multiplying both the top and bottom by the same factor.

$$\frac{3}{24} = \frac{3 \times 7}{24 \times 7} = \frac{21}{168}$$

$$\frac{5}{56} = \frac{5 \times 3}{56 \times 3} = \frac{15}{168}$$

Thus, $\frac{3}{24} + \frac{5}{56} = \frac{21}{168} + \frac{15}{168} = \frac{36}{168}$. Since the greatest common factor of 36 and 168 is 12, $\frac{36}{168}$ simplifies to $\frac{3}{14}$.

Multiplying Fractions

To multiply fractions, multiply the numerators and then multiply the denominators. Simplify if necessary. It may be possible to cross simplify prior to multiplying.

$$\text{First Method:} \quad \frac{3}{4} \times \frac{2}{9} = \frac{3 \times 2}{4 \times 9} = \frac{6}{36} = \frac{1}{6}$$

$$\text{Second Method:} \quad \frac{3}{4} \times \frac{2}{9} = \frac{3}{9} \times \frac{2}{4} = \frac{1}{3} \times \frac{1}{2} = \frac{1}{6}$$

Dividing Fractions

To divide fractions, multiply by the reciprocal, or multiplicative inverse, of the divisor (second fraction). When you multiply a number by its respective multiplicative inverse, or reciprocal, the product is 1. For example, the reciprocal of 4 is $\frac{1}{4}$ because $4 \times \frac{1}{4} = 1$. Observe the example below. The divisor, or the number you are dividing by, is the second fraction $\frac{2}{7}$. Thus, multiply $\frac{1}{5}$ by the reciprocal of $\frac{2}{7}$. The reciprocal of $\frac{2}{7}$ is $\frac{7}{2}$.

$$\frac{1}{5} \div \frac{2}{7} = \frac{1}{5} \times \frac{7}{2} = \frac{7}{10}$$

Converting Fractions, Decimals, and Percentages

In order to **convert fractions into decimals**, simply divide. However, it is necessary to memorize some common fractions as depicted by the table below.

$\frac{1}{10} = 0.1$	$\frac{2}{10} = \frac{1}{5} = 0.2$	$\frac{3}{10} = 0.3$	$\frac{4}{10} = \frac{2}{5} = 0.4$	$\frac{5}{10} = \frac{1}{2} = 0.5$
$\frac{6}{10} = \frac{3}{5} = 0.6$	$\frac{7}{10} = 0.7$	$\frac{8}{10} = \frac{4}{5} = 0.8$	$\frac{9}{10} = 0.9$	$\frac{10}{10} = 1$

$\frac{1}{9} = 0.111\cdots = 0.\overline{1}$	$\frac{2}{9} = 0.222\cdots = 0.\overline{2}$	$\frac{3}{9} = 0.333\cdots = 0.\overline{3}$	$\frac{4}{9} = 0.444\cdots = 0.\overline{4}$
$\frac{5}{9} = 0.555\cdots = 0.\overline{5}$	$\frac{6}{9} = 0.666\cdots = 0.\overline{6}$	$\cdots$	$\frac{9}{9} = 1$

$\frac{1}{6} = 0.166\cdots = 0.1\overline{6}$	$\frac{2}{6} = \frac{1}{3} = 0.333\cdots = 0.\overline{3}$	$\frac{4}{6} = \frac{2}{3} = 0.666\cdots = 0.\overline{6}$	$\frac{5}{6} = 0.833\cdots = 0.8\overline{3}$

$\frac{1}{8} = 0.125$	$\frac{2}{8} = \frac{1}{4} = 0.25$	$\frac{3}{8} = 0.375$	$\frac{4}{8} = \frac{1}{2} = 0.5$
$\frac{5}{8} = 0.625$	$\frac{6}{8} = \frac{3}{4} = 0.75$	$\frac{7}{8} = 0.875$	$\frac{8}{8} = 1$

When **converting fractions into percentages**, attempt to set the denominator to 100 by multiplication. After multiplying both the numerator and denominator by the same factor, the numerator represents the percentage.

$$\frac{4}{25} = \frac{4 \times 4}{25 \times 4} = \frac{16}{100} = 16\%$$

If it is not possible to convert the denominator into 100 or deemed too complicated, convert the fraction into a decimal by dividing and then convert the decimal into percentage.

To **convert decimals into percentages**, multiply the number by 100 and add the % (percent) sign or move the decimal point two places to the right. In order to **convert percentages into decimals**, divide the number by 100 and remove the % (percent) sign or move the decimal point two places to the left.

Integers

Natural numbers are positive counting numbers starting from one: $1, 2, 3, \cdots, +\infty$. **Whole numbers** are counting numbers including zero: $0, 1, 2, \cdots, +\infty$. A hint to remember that whole numbers include zero is that the word "wh**o**le" has the letter o which looks like the number zero, 0. **Integers** are counting numbers which include zero, and negative and positive numbers: $-\infty, \cdots, -2, -1, 0, 1, 2, \cdots, +\infty$.

Since integers consist of both positive and negative numbers, there are different rules when performing the four operations: adding, subtracting, multiplying, and dividing. When evaluating an expression, use the **order of operations (PEMDAS)**. The order of operations suggests to first perform any calculations inside parentheses. Afterwards, evaluate any exponents. Next, perform all multiplications and divisions working from left to right. Finally, do additions and subtractions from left to right.

When **adding integers**, you must take into consideration the magnitude of the positive and negative values. When adding two positive integers, the sum becomes more positive. Likewise, when adding two negative integers, the sum becomes more negative. Let's use a common metaphor which relates positive and negative values. Think of positive integers as good guys and negative integers as bad guys. Good guys plus good guys will equal even more good guys. Likewise, bad guys plus bad guys result in even more bad guys. When adding a positive integer and a negative integer, compare to see which of the following has more value. For example, 5 good guys verses 2 bad guys will result in a win for the good guys because there are 3 more good guys. Thus, when adding $5 + (-2)$, the sum is positive 3. On the other hand, if 5 bad guys verses 2 good guys, the bad guys will win because they have a stronger pull by 3 bad guys. Since there are 3 bad guys remaining and bad guys are negative, when adding $-5 + 2$, the sum is -3. The metaphor, although simple, proves the theory of adding integers. Be careful and pay close attention to which number, positive or negative, has a larger impact.

Another concept to understand is that adding a negative is the same concept as subtracting. Let's use $6 + (-4)$ for example. Since adding a negative means subtraction, $6 + (-4)$ can be rewritten as $6 - 4 = 2$. Even if the question was originally written $(-4) + 6$, it is possible to use the commutative property of addition to rewrite it as $6 + (-4)$.

$$6 + 4 = 10 \qquad \text{(Adding two positive integers)}$$
$$-7 + (-5) = -12 \qquad \text{(Adding two negative integers)}$$
$$7 + (-18) = -11 \qquad \text{(Adding a positive and negative integer)}$$
$$(-4) + 7 = 3 \qquad \text{(Adding a positive and negative integer)}$$

Pay close attention when **subtracting integers**. When subtracting integers, picture a number line. For example, let's take the problem $6 - 8$. Since 6 is positive, walk 6 steps right from the starting position. Afterwards, since you are subtracting 8 from the new location, walk left 8 spaces and you will be 2 spaces left from the starting position. Thus, $6 - 8 = -2$. The only tricky aspect of subtracting integers is when subtracting a negative number. In the example, $-4 - (-8)$, we start off by walking 4 steps left from the starting position. Although subtracting indicates walking left, subtracting a negative reverses the walking direction and instead you must walk 8 steps to the right. In other words, subtracting a negative number means to add. Therefore, $-4 - (-8)$ can be rewritten as $-4 + 8$ which equals positive 4. Adding and subtracting integers can be placed into two rules. When two like signs are adjacent to each other, it means to add. When two unlike signs are adjacent to each other, it means to subtract.

In order to **multiply integers**, the number of the product will be the same but it is necessary to observe whether or not the number is positive or negative. When multiplying two like signs together, the answer is always positive: $+ \times + = +$ and $- \times - = +$. When multiplying a positive and negative number, the product is always negative: $+ \times - = -$ and $- \times + = -$. When an expression has only multiplication, it is possible to determine whether or not the product is positive or negative by counting the number of negative terms. If the expression has an even number of negative values, the product will be positive. However, if the expression consists of an odd number of negative values, the product will be negative. It is also possible to take each term in the expression step by step. For example, $-4 \times 4 \times (-4) = -16 \times (-4) = +64$.

$$5 \times 4 = 20 \quad \text{(Multiplying two positive integers)}$$
$$-5 \times (-4) = 20 \quad \text{(Multiplying two negative integers)}$$
$$-5 \times 4 = -20 \quad \text{(Multiplying a positive and negative integer)}$$
$$5 \times (-4) = -20 \quad \text{(Multiplying a positive and negative integer)}$$

$$2 \times (-2) \times (-2) \times (-2) \times (-2) = 32 \quad \text{(Product is positive because even number of negative integers)}$$
$$2 \times (-2) \times (-2) \times (-2) \times 2 = -32 \quad \text{(Product is negative because odd number of negative integers)}$$

Dividing integers follow the same rules as multiplying integers. When dividing two like signs together, the answer is always positive: $+ \div + = +$ and $- \div - = +$. When dividing a positive and negative number, the quotient is always negative: $+ \div - = -$ and $- \div + = -$.

$$20 \div 5 = 4 \quad \text{(Dividing two positive integers)}$$
$$-20 \div (-5) = 4 \quad \text{(Dividing two negative integers)}$$
$$-20 \div 5 = -4 \quad \text{(Dividing a positive and negative integer)}$$
$$20 \div (-5) = -4 \quad \text{(Dividing a positive and negative integer)}$$

Algebraic Properties

Commutative Property of Addition: The order in which numbers are added in an expression does not change the sum: $x + y = y + x$.

$$5 + 4 = 4 + 5$$
$$9 = 9$$

Commutative Property of Multiplication: The order in which numbers are multiplied in an expression does not change the product: $x \cdot y = y \cdot x$.

$$5 \times 4 = 4 \times 5$$
$$20 = 20$$

Associative Property of Addition: When adding, the order and way in which numbers are grouped does not change the sum: $(x + y) + z = x + (y + z)$.

$$(3 + 4) + 5 = 3 + (4 + 5)$$
$$7 + 5 = 3 + 9$$
$$12 = 12$$

Associative Property of Multiplication: When multiplying, the order and way in which numbers are grouped does not change the product: $(x \times y) \times z = x \times (y \times z)$.

$$(3 \cdot 4) \cdot 5 = 3 \cdot (4 \cdot 5)$$
$$12 \cdot 5 = 3 \cdot 20$$
$$60 = 60$$

Additive Identity: When adding 0 to any number, the sum is the number: $x + 0 = x$.

$$-10 + 0 = -10$$

Multiplicative Identity: When multiplying 1 to any number, the product is the number: $x \cdot 1 = x$.

$$-10 \cdot 1 = -10$$

Multiplicative Property of Zero: When multiplying 0 to any number, the product is zero: $x \cdot 0 = 0$.

$$-10 \cdot 0 = 0$$

Symmetric Property: If one measure equals a second measure, then the second measure equals the first: if $a = b$, then $b = a$.

$$\text{If } 4 + 3 = 7, \text{ then } 7 = 4 + 3.$$

Transitive Property: If one measure equals a second measure and the second measure equals a third measure, then the first measure equals the third measure: if $a = b$ and $b = c$, then $a = c$.

$$\text{If } 2 + 7 = 3 \times 3 \text{ and } 3 \times 3 = 9, \text{ then } 2 + 7 = 9.$$

Distributive Property: The distributive property demonstrates multiplying a number by a sum. In other words, multiply each number inside a parenthesis by the number outside of the parenthesis. When evaluating expressions, it is necessary to solve expressions inside the parentheses first; however, the distributive property is necessary when dealing with variables inside the parenthesis.

$$2(3 + 5) = 2(3) + 2(5)$$

As illustrated in the example above, the 2 is distributed to the 3 and then the 2 is distributed to the 5. Afterwards, proceed to multiply and add.

$$2(3) + 2(5) = 6 + 10 = 16$$

The number you are multiplying by can be written outside of the parenthesis either on the left or right side: $a(b+c)$ or $(b+c)a$. Both are correct; however, $a(b+c)$ is more commonly represented in algebraic problems. It is necessary to understand how negative numbers are distributed. For example, $10 - (2 + 3)$ is not the same as $10 - 2 + 3$. The negative sign must be distributed into both terms inside the parenthesis: $10 - 2 - 3$. Likewise, when dealing with negative numbers, the negative must be distributed into each term within the parenthesis. The eight examples below represent the different possible outcomes when multiplying a term and a binomial.

$$a(b + c) = ab + ac$$
$$a(b - c) = ab - ac$$
$$a(-b + c) = -ab + ac$$
$$a(-b - c) = -ab - ac$$
$$-a(b + c) = -ab - ac$$
$$-a(b - c) = -ab + ac$$
$$-a(-b + c) = ab - ac$$
$$-a(-b - c) = ab + ac$$

Variables

Variables are letters that act as an unknown in a problem and are depicted by a variety of letters including, but not limited to, x, y, and n. In simple terms, observe the equation $x+2=5$. Since $3+2=5$, the variable x is equal to 3 as it serves as a placeholder and unknown in the problem.

Combining Like Terms

Like terms are terms that have same variables and same exponent; only the coefficients may be different but can be the same. Knowing like terms is essential when you simplify algebraic expressions. For instance,

- $2x$ and $3x$: (Like terms)
- $2x$ and $3x^2$: (Not like terms since the two expressions have different exponents)
- 2 and 3: (Like terms)

To simplify algebraic expressions, expand the expression using the distributive property when necessary. Then group the like terms and simplify them. For instance,

$$\begin{aligned} 2(-x+2)+3x+5 &= -2x+4+3x+5 \quad &\text{(Use distributive property to expand)} \\ &= (-2x+3x)+(4+5) \quad &\text{(Group the like terms and simplify)} \\ &= x+9 \end{aligned}$$

To evaluate an algebraic expression, substitute the numerical value into the variable. When substituting a negative numerical value, make sure to use a **parenthesis** to avoid a mistake.

Solving Equations and Word Problems

Solving an equation is finding the value of the variable that makes the equation true. In order to solve an equation, use the rule called SADMEP with inverse operations (SADMEP is the reverse order of the order of operations, PEMDAS). Inverse operations are the operations that cancel each other. Addition and subtraction, and multiplication and division are good examples.

SADMEP suggests to first cancel subtraction or addition. Then, cancel division or multiplication next by applying corresponding inverse operation. Below is an example that shows you how to solve $2x - 1 = 5$, which involves subtraction and multiplication.

$$\begin{aligned} 2x - 1 &= 5 \\ +1 &= +1 \\ 2x &= 6 \\ x &= 3 \end{aligned}$$

$\checkmark \quad \checkmark$
$S\ A\ D\ M\ E\ P$
Addition to cancel substraction
Division to cancel multiplication

Solving word problems involve translating verbal phrases into mathematical equations. The table below summarizes the guidelines.

Verbal Phrase	Expression
A number	x
Is	$=$
Of	$\times$
Percent	0.01 or $\frac{1}{100}$
The sum of x and y	$x + y$
Three more than twice a number	$2x + 3$
The difference of x and y	$x - y$
3 is subtracted from a number	$x - 3$
4 less than a number	$x - 4$
A number decreased by 5	$x - 5$
6 less a number	$6 - x$
The product of x and y	xy
6 times a number	$6x$
The quotient of x and y	$\frac{x}{y}$
A number divided by 9	$\frac{x}{9}$

For instance, use SADMEP to solve for x in the verbal phrase: 5 more than the quotient of x and 3 is 14.

$$\frac{x}{3} + 5 = 14 \qquad \checkmark\checkmark$$
$$\phantom{\frac{x}{3}}\ \ S\ A\ D\ M\ E\ P$$

$\frac{x}{3} + 5 = 14$ $S\ A\ D\ M\ E\ P$

$-5 = -5$ Subtraction to cancel addition

$\frac{x}{3} = 9$ Multiplication to cancel division

$x = 27$

Solving Inequalities

Solving an inequality is exactly the same as solving an equation. To solve an inequality, use SADMEP (Reverse order of the PEMDAS). In most cases, the inequality symbol remains unchanged. However, there are only two cases in which the inequality symbol must be reversed. The first case is when you multiply or divide each side by a negative number. The second case is when you take a reciprocal of each side. For instance,

$$
\begin{array}{ll}
\underline{\text{Case 1}} & \underline{\text{Case 2}} \\
2 < 3 & 2 < 3 \\
-2 > -3 & \dfrac{1}{2} > \dfrac{1}{3}
\end{array}
$$

For instance, use SADMEP to solve the inequality $-3x + 2 > x + 10$.

$$
\begin{aligned}
-3x + 2 &> x + 10 & &\text{Subtract } x \text{ from each side} \\
-4x + 2 &> 10 & &\text{Subtract 2 from each side} \\
-4x &> 8 & &\text{Divide each side by } -4 \\
x &< -2 & &\text{Reverse the inequality symbol}
\end{aligned}
$$

Ratios, Rates, and Proportions

A **ratio** is a fraction that compares two quantities measured in the same units. The ratio of a to b can be written as $a : b$ or $\frac{a}{b}$. If the ratio of a number of apples to that of oranges in a store is $3 : 4$ or $\frac{3}{4}$, it means that there are 3 apples to every 4 oranges in the store.

A **rate** is a ratio that compares two quantities measured in different units. A rate is usually expressed as a unit rate. A unit rate is a rate per one unit of a given quantity. The rate of a per b can be written as $\frac{a}{b}$. If a car travels 100 miles in 2 hours, the car travels at a rate of 50 miles per hour.

A **proportions** is an equation that states that two ratios are equal A proportion can be written as

$$a : b = c : d \quad \text{or} \quad \frac{a}{b} = \frac{c}{d}$$

The proportion above reads a is to b as c is to d. To solve the value of a variable in a proportion, use the cross product property and then solve for the variable. For instance,

$$
\begin{aligned}
\frac{x}{2} &= \frac{6}{3} & &\text{Cross Product Property} \\
3x &= 2 \times 6 \\
x &= 4
\end{aligned}
$$

Properties of Exponents

In the expression 2^4, 2 is the base, 4 is the exponent, and 2^4 is the power. Exponents represent how many times the base is multiplied by. $2^4 = 2 \times 2 \times 2 \times 2$. The table below shows a summary of the properties of exponents .

Properties of Exponents	Example
1. $a^m \cdot a^n = a^{m+n}$	1. $2^4 \cdot 2^6 = 2^{10}$
2. $\frac{a^m}{a^n} = a^{m-n}$	2. $\frac{2^{10}}{2^3} = 2^{10-3} = 2^7$
3. $a^0 = 1$	4. $(-2)^0 = 1, (3)^0 = 1, (100)^0 = 1$
4. $a^{-1} = \frac{1}{a}$	5. $2^{-1} = \frac{1}{2}$

Scientific Notation

Scientific notation is the process of writing large or small numbers in the form of $c \times 10^n$ where $1 \leq c < 10$ and n is an integer. In other words, c must be greater than or equal to 1 but less than 10. In general, positive n values give a large number while negative n values produce small fractional values. It is important to understand how to convert numbers written in standard notation to scientific notation and vice versa.

To **convert standard notation to scientific notation**, set the decimal point to create a number that satisfies the definition of c. For example, the c value of 12,400 is 1.24. Since 12,400 is a large number and the decimal point moved 4 places to the left, the scientific notation is written as $12,400 = 1.24 \times 10,000 = 1.24 \times 10^4$. For smaller numbers such as 0.000000024, the c value is 2.4. Since the number is small and the decimal point moved 8 places to the right, the scientific notation is written as $0.000000024 = \frac{2.4}{100,000,000} = 2.4 \times 10^{-8}$

Functions and Relations

An ordered pair (x, y) describes the location of a point on the coordinate system. The first number in the ordered pair describes the x-coordinate, or the position relevant to the x-axis. The x-coordinate is often referred to as the **input** and the set of all x-values in a given set is called the **domain**. The second number in the ordered pair describes the y-coordinate, or the position relevant to the y-axis. The y-coordinate is often referred to as the **output** and the set of all y-values in a given set is called the **range**. A set of ordered pairs is described as a **relation**. A relation can be depicted in several ways.

$$\{(1, 2), (2, 3), (3, 4), (4, 5)\}$$

x	1	2	3	4
y	2	3	4	5

x	y
1	2
2	3
3	4
4	5

A **function** is a special type of relation that relates a specific input to exactly one output. Each member of the domain, or x-value, is paired with only one member of the range, or y-value. The x-value cannot repeat nor have multiple y-coordinates. The y-value can repeat. In other words, a function relates an input to an output so that an input x cannot have more than one value for an output y. For instance, let's consider the following scenario. On Monday, you make 10 dollars. Is it, then, possible to make 10 dollars again on Tuesday? Sure it is. There is nothing stopping you from making the same about of money you did on the previous day. This shows the concept of how the y-coordinate can repeat which represents the dollars earned in a given day of the week. However, this next statement shows ambiguity and is unclear on what it is trying to display. On Monday, you make 10 dollars. On Monday, you make 5 dollars. Did you make 5 dollars or 10 dollars on Monday. If you made 15 dollars, why didn't you state, "On Monday, I made 15 dollars." This is an example of why a function defines a relation that every member of the domain is paired with exactly one member of its range. If you made 5 dollars on the Monday of Week 1 and 10 dollars on the Monday of Week 2, then those are two separate x-coordinate values: $(Mon_1, 5), (Mon_2, 10)$.

When illustrated as a graph or scatter plot, it is possible to determine whether or not the shown relation is a function or not. Use the **vertical line test**. If, at any position on the graph, a vertical line can be drawn that passes through two different points, the relation fails the vertical line test and is **NOT** a function.

A function can be described using a symbol. For example, $a \odot b = \frac{2a}{b}$. The symbol, $\odot$, represents the expression $\frac{2a}{b}$ as where a and b are both variables. Thus, if a and b are given, it is possible to evaluate the expression. If $a \odot b = \frac{2a}{b}$, what is $3 \odot 9$?

$$a \odot b = \frac{2a}{b}$$

$$3 \odot 9 = \frac{2(3)}{9} = \frac{6}{9} = \frac{2}{3}$$

Linear Equations

The **slope**, m, of a line is a number that describes the steepness of the line. If a line passes through the points (x_1, y_1) and (x_2, y_2), the slope m is defined as

$$m = \frac{\text{Rise}}{\text{Run}} = \frac{y_2 - y_1}{x_2 - x_1}$$

An equation of a line can be written in slope-intercept form, $y = mx + b$, where m is slope and b is y-intercept. Below classifies the lines by slope.

- Lines that rise from left to right have positive slope.
- Lines that fall from left to right have negative slope.
- Horizontal lines have zero slope (example: $y = 2$).
- Vertical lines have undefined slope (example: $x = 2$).
- Parallel lines have the same slope.

The x-**intercept** of a line is a point where the line crosses x-axis. The y-**intercept** of a line is a point where the line crosses y-axis.

| To find the x-intercept of a line | $\implies$ | Substitute 0 for y and solve for x |
| To find the y-intercept of a line | $\implies$ | Substitute 0 for x and solve for y |

IAAT PRACTICE TEST 1

SECTION 1
Time — 10 minutes
15 Questions

Directions: Read the information given and choose the best answer for each question. Base your answer only on the information given. The time limit for each section is 10 minutes.

1. $35 \times 0.25 =$

   (A) 0.875
   (B) 8.75
   (C) 87.5
   (D) 875

2. $3.8 + 234 + 15.3 + 68 =$

   (J) 311.1
   (K) 321.1
   (L) 330.1
   (M) 331.1

3. $34.1 \div 15.5 =$

   (A) 0.46
   (B) 1.2
   (C) 1.8
   (D) 2.2

4. $1354 - 856 =$

   (J) 478
   (K) 488
   (L) 498
   (M) 508

5. Simplify $2\frac{1}{3} + \frac{5}{3}$.

   (A) 4
   (B) $3\frac{2}{3}$
   (C) 3
   (D) $2\frac{2}{3}$

6. If the weight of a school bus is 14,000 pounds, what is 14,000 written in scientific notation?

   (J) 14×10^{-4}
   (K) 1.4×10^{-3}
   (L) 14×10^{3}
   (M) 1.4×10^{4}

7. A water cooler contains $2\frac{3}{4}$ gallons of drinking water. After one day, $\frac{7}{8}$ gallon of water had been used. How much water was left in the water cooler?

   (A) $2\frac{1}{8}$
   (B) $1\frac{7}{8}$
   (C) $1\frac{5}{8}$
   (D) $1\frac{3}{8}$

8. A sales tax rate is 8%. If Joshua wants to buy a bicycle which is $250, how much does Joshua need to pay including sales tax?

   (J) $270
   (K) $265
   (L) $260
   (M) $255

9. Jason mixes paint using 7 ounces of white paint for every 3 ounces of green paint. At this rate, how many ounces of green paint would be mixed with 28 ounces of white paint?

   (A) 15
   (B) 14
   (C) 13
   (D) 12

10. Which of the following set is ordered from least to greatest?

    (J) $\{0.8, 20\%, \frac{1}{4}, 1.5\}$
    (K) $\{1.5, 0.8, \frac{1}{4}, 20\%\}$
    (L) $\{20\%, \frac{1}{4}, 0.8, 1.5\}$
    (M) $\{\frac{1}{4}, 0.8, 1.5, 20\%\}$

11. Which of the following correctly converts $\frac{1}{8}$ to a percent?

    (A) 125%
    (B) 12.5%
    (C) 1.25%
    (D) 0.125%

12. A baseball team wins 3 games every 4 games played. If the baseball team played 60 games, how many games did it lose?

    (J) 10
    (K) 15
    (L) 20
    (M) 45

13. What is the remainder when 111 is divided by 17?

    (A) 6
    (B) 7
    (C) 8
    (D) 9

14. There are forty eight candies in a box. If Joshua eats half of the candies and then Jason eats three-fourth of the remaining candies, how many candies are left in the box?

   (J) 10
   (K) 8
   (L) 6
   (M) 4

15. If a $1200 computer is discounted 25% off, what is the new price of the computer?

   (A) $700
   (B) $800
   (C) $900
   (D) $1000

STOP

IAAT PRACTICE TEST 1

SECTION 2
Time — 10 minutes
15 Questions

Directions: Read the information given and choose the best answer for each question. Base your answer only on the information given. The time limit for each section is 10 minutes.

Directions: Use the following graph to answer questions 1 – 4.

Fairfax County Athlete Participation

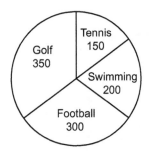

Golf 350
Tennis 150
Swimming 200
Football 300

1. What is the total number of athletes in Fairfax County?

    (A) 850
    (B) 900
    (C) 950
    (D) 1000

2. What percentage of Fairfax County athletes participated in tennis?

    (J) 35%
    (K) 25%
    (L) 20%
    (M) 15%

3. How many more athletes participated in the most favorite sport than the least favorite sport?

    (A) 100
    (B) 150
    (C) 200
    (D) 250

4. According to the pie chart, what is the probability that an athlete chosen at random participates in football?

    (J) $\frac{1}{5}$
    (K) $\frac{3}{10}$
    (L) $\frac{1}{3}$
    (M) $\frac{2}{5}$

Directions: Use the following graph to answer questions 5 – 8.

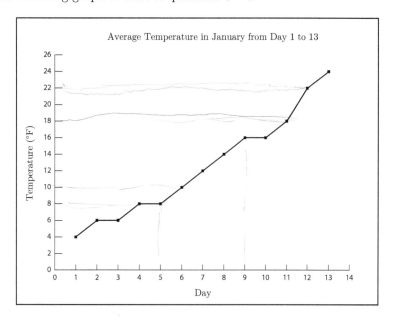

5. Which statement best describes the temperatures in January from day 1 to 13?

   (A) The temperatures increased gradually throughout this period.

   (B) The temperatures decreased until day 8 and then increased rapidly.

   (C) The temperatures increased until day 9 and then held fairly constant.

   (D) The temperatures held fairly constant until day 5 and then decreased at a steady rate.

6. What day in January was the last time the temperature did **NOT** increase?

   (J) From day 3 to day 4

   (K) From day 7 to day 8

   (L) From day 9 to day 10

   (M) From day 12 to day 13

7. How much did the temperature increase per day between day 5 and day 9?

   (A) $1°F$ per day

   (B) $2°F$ per day

   (C) $3°F$ per day

   (D) $4°F$ per day

8. What was the largest increase of temperature for one day in January?

   (J) $8°F$

   (K) $6°F$

   (L) $4°F$

   (M) $2°F$

Item	Cost
Eggs	$1.50
Milk	$3.50
Flour	$2.00
Soda	$3.25

9. Sue went to the store to purchase the following items shown above. If Sue gave the cashier $20, how much did Sue receive back in change?

   (A) $9.25
   (B) $9.75
   (C) $10.25
   (D) $11.75

M	T	W	R	F	Total
2.5	4	3	2.4	1.3	

11. The table above shows the distance, in miles, that Mr. Rhee ran each day. What is the total distance, in miles, that Mr. Rhee ran in the five days?

    (A) 14.9 miles
    (B) 14.3 miles
    (C) 13.6 miles
    (D) 13.2 miles

♣, ♦, ♡, ♠, ♣, ♦, ⋯

10. If the pattern shown above continues, what is the 15th shape of the pattern?

    (J) ♣
    (K) ♦
    (L) ♡
    (M) ♠

Directions: Use the following table to answer questions 12 – 14.

Name	Game 1	Game 2	Game 3	Game 4
Alex	1	0	1	3
Joshua	0	3	0	5
Jason	3	4	4	1

12. Who did score the least for the four games?

    (J) Alex
    (K) Joshua
    (L) Jason
    (M) There is not enough information given to answer this question

13. What is Jason's average score for the four games?

    (A) 3
    (B) 2
    (C) 1
    (D) 0

14. On which of the game is the sum of the three scores the largest?

   (J) Game 1
   (K) Game 2
   (L) Game 3
   (M) Game 4

Day	Temperature
Monday	52
Tuesday	52
Wednesday	60
Thursday	70
Friday	76

15. The table above shows the temperatures for Fairfax City. What is the mean (average) of the temperatures?

   (A) 58
   (B) 60
   (C) 62
   (D) 64

STOP

IAAT PRACTICE TEST 1

SECTION 3
Time — 10 minutes
15 Questions

Directions: Read the information given and choose the best answer for each question. Base your answer only on the information given. The time limit for each section is 10 minutes.

1. Which of the following is **NOT** a function?

    (A) $\{(5,4),(3,5),(2,9),(8,5)\}$
    (B) $\{(1,2),(3,2),(5,2),(7,2)\}$
    (C) $\{(3,2),(2,4),(5,-1),(3,-2)\}$
    (D) $\{(5,4),(1,2),(3,9),(2,8)\}$

2. What is the slope of the line that passes through the points $(-1,-2)$ and $(1,2)$?

    (J) 1
    (K) 2
    (L) 3
    (M) 4

x	1	2	3
y	2	4	6

3. The table above shows three pairs of x and y values. Which of the following equation is true for all values in the table?

    (A) $y = x$
    (B) $y = 2x$
    (C) $y = 3x$
    (D) $y = 4x$

x	y
1	3
2	3
3	3

4. The table above shows three ordered pairs. Which of the following equation best represents the three ordered pairs?

    (J) $y = x + 2$
    (K) $y = 3x$
    (L) $y = 3$
    (M) $x = 3$

5. Evaluate $y = \frac{1}{2}(1-x)$ when $x = \frac{1}{3}$.

    (A) $\frac{3}{4}$
    (B) $\frac{2}{3}$
    (C) $\frac{1}{2}$
    (D) $\frac{1}{3}$

6. If $y = 3x + 4$ and $y = 13$, what is the value of x?

    (J) 3
    (K) 2
    (L) 1
    (M) 0

Input	Output
2	5
3	7
4	9
5	

7. Observe the numbers in the two columns shown above. Which of the following value should be in the empty cell?

   (A) 10
   (B) 11
   (C) 12
   (D) 13

Input	Output
1	4
2	7
3	10
⋮	⋮
	19

8. Observe the numbers in the two columns shown above. Which of the following value should be in the empty cell?

   (J) 7
   (K) 6
   (L) 5
   (M) 4

9. Which of the following table contains only ordered pairs that satisfy the equation, $y = 2x - 2$?

(A)
x	1	2	3	4
y	0	3	6	9

(B)
x	-2	-1	0	1
y	2	0	-2	-4

(C)
x	0	1	2	3
y	0	2	4	6

(D)
x	1	2	3	4
y	0	2	4	6

Expenses (E)	Profit (P)
50	65
100	130
150	195
200	260

10. The table above shows the expenses, E, and profit, P, of a company. Which of the following equation best represents the relationship between the expenses and profit of the company?

   (J) $P = 0.7E$
   (K) $P = 0.9E$
   (L) $P = 1.1E$
   (M) $P = 1.3E$

11. Which of the following table best represents the following verbal relationship? The number of burgers sold, y, is four more than three times the number of sodas sold, x.

(A)
x	2	3	4	5
y	10	13	16	19

(B)
x	1	2	3	4
y	5	8	11	14

(C)
x	0	1	2	3
y	4	9	14	19

(D)
x	1	2	3	4
y	-1	2	5	8

12. Which of the following equation best represents the following verbal relationship? The number of girls, g, is five less than four times the number of boys, b.

   (J) $g = 4b - 5$
   (K) $b = 4g - 5$
   (L) $g = 5 - 4b$
   (M) $b = 5 - 4g$

13. Which of the following equation represents a line that is parallel to the following equation, $y = -3x + 4$?

   (A) $y = \frac{1}{3}x + 3$
   (B) $y = 3x - 4$
   (C) $y = -3x + 8$
   (D) $y = -4x - 4$

14. A linear equation $y = 2x + 6$ represents a straight line in the xy coordinate plane. What is the y-intercept of the line?

   (J) 6
   (K) 4
   (L) 2
   (M) 1

Quantity	10	20	40	80	160
Unit Price	$10	$9.5	$9	$8.5	$8

15. A t-shirt company provides lower prices if shirts are purchased in larger quantities. The table above provides their prices. If the amount of quantity reaches 640, what will be the unit price?

   (A) $7.5
   (B) $7
   (C) $6.5
   (D) $6

STOP

IAAT PRACTICE TEST 1

SECTION 4
Time — 10 minutes
15 Questions

Directions: Read the information given and choose the best answer for each question. Base your answer only on the information given. The time limit for each section is 10 minutes.

1. In the figure below, the two rectangles are similar.

    Which proportion can be used to find x?

    (A) $\dfrac{10}{15} = \dfrac{x}{6}$

    (B) $\dfrac{15}{10} = \dfrac{6}{x}$

    (C) $\dfrac{10}{6} = \dfrac{15}{x}$

    (D) $\dfrac{10}{6} = \dfrac{x}{15}$

2. If $x = 3$ and $y = -4$, what is the value of $2x - y$?

    (J) 2
    (K) 6
    (L) 10
    (M) 12

3. Solve for x: $\dfrac{3}{8} = \dfrac{12}{x}$

    (A) 32
    (B) 36
    (C) 40
    (D) 42

4. $-2(3x - 4) =$

    (J) $6x + 8$
    (K) $6x - 8$
    (L) $-6x + 8$
    (M) $-6x - 4$

5. If $x = -6$ and $y = -12$, which of the following expression has the largest value?

    (A) $\dfrac{y}{x}$
    (B) xy
    (C) $x + y$
    (D) $x - y$

6. Solve for x: $3x + 2 = 17$

   (J) 5
   (K) 6
   (L) 7
   (M) 8

7. Which of the following value satisfies the inequality $11 - x < -4$?

   (A) 13
   (B) 14
   (C) 15
   (D) 16

8. Jason can paint 6 pictures in 3 hours. Which proportion can be used to find p, the number of pictures Jason can paint in 7 hours?

   (J) $\frac{6}{7} = \frac{3}{p}$
   (K) $\frac{6}{3} = \frac{p}{7}$
   (L) $\frac{p}{6} = \frac{3}{7}$
   (M) $\frac{7}{3} = \frac{6}{p}$

9. The volume of a cylinder is defined as $V = \pi r^2 h$, where r and h are the radius and height, respectively. What is the volume of a cylinder that has a height of 7 and a diameter of 4?

   (A) 21π
   (B) 28π
   (C) 35π
   (D) 112π

10. If $x + 2y = z$ and $z = 6 + x$, what is the value of y ?

    (J) 12
    (K) 6
    (L) 4
    (M) 3

11. If $a \blacktriangle b = \dfrac{a+b}{a-b}$, what is the value of $6 \blacktriangle 3$?

    (A) $\frac{1}{3}$
    (B) $\frac{1}{2}$
    (C) 2
    (D) 3

12. In the equation $x + y = z$, if x is increased by 3 and y is decreased by 8, what is the new value of z ?

    (J) $z - 5$
    (K) $z - 3$
    (L) $z + 3$
    (M) $z + 5$

13. If $x = 3$ and $3x + 4y = 25$, what is the value of y ?

    (A) 1
    (B) 2
    (C) 4
    (D) 6

14. Which phrase is represented by $2(4+x)+3$?

(J) The sum of three and two times the difference between a number, x, and four.

(K) Three more than two times the sum of four and a number, x.

(L) Three more than the sum of two times a number, x, and four.

(M) The sum of three and two times the quotient of four and a number, x.

15. Simplify the expression $4x + 3x + 2$.

(A) $14x + 2$

(B) $12x + 2$

(C) $9x + 2$

(D) $7x + 2$

STOP

SOLOMON ACADEMY — TEST 1 SECTION 1

Answers and Solutions
IAAT Practice Test 1 Section 1

Answers

1. B	2. K	3. D	4. L	5. A
6. M	7. B	8. J	9. D	10. L
11. B	12. K	13. D	14. L	15. C

Solutions

1. (B)

    In order to multiply decimal numbers, multiply normally, ignoring the decimal points first. Afterwards, place the decimal point into the solution equivalent to the amount of decimal places the two numbers have together. For example, change 35×0.25 into $35 \times 25 = 875$. Since 35 has no decimal places and 0.25 has two numbers after the decimal point, the solution 875 will contain two decimal places. Therefore, $35 \times 0.25 = 8.75$.

2. (K)

    When adding decimal numbers, it is necessary to line up the decimal points prior to adding. Make sure to add properly and carry over any necessary digits.

    $$\begin{array}{r} 3.8 \\ 234.0 \\ 15.3 \\ +68.0 \\ \hline 321.1 \end{array}$$

3. (D)

    When dividing decimal numbers, it is necessary to convert the number you are dividing by into a whole number. Convert 34.1 and 15.5 into 341 and 155 respectively by shifting the decimal point of both numbers one place to the right. Afterwards, perform long division.

    $$\frac{34.1}{15.5} = \frac{341}{155} = 2.2$$

4. (L)

    When subtracting, line up the digits place and do not forget to borrow when necessary.

    $$1354 - 856 = 498$$

5. **(A)**

When adding fractions, make sure that the denominator is the same. Note that the mixed number, $2\frac{1}{3}$, can be expressed as $2 + \frac{1}{3}$.

$$2\frac{1}{3} + \frac{5}{3} = 2 + \left(\frac{1}{3} + \frac{5}{3}\right)$$
$$= 2 + \left(\frac{6}{3}\right)$$
$$= 4$$

Additionally, it is also possible to add two fractions by converting the mixed number into an improper fraction.

$$2\frac{1}{3} + \frac{5}{3} = \frac{7}{3} + \frac{5}{3} = \frac{12}{3} = 4$$

6. **(M)**

In scientific notation, all the numbers can be written in the form of $c \times 10^n$, where $1 \leq c < 10$ and n is an integer.

$$14000 = 1.4 \times 10000 = 1.4 \times 10^4$$

Therefore, 14,000 written in scientific notation is 1.4×10^4.

7. **(B)**

In order to figure out how much water was left in the water cooler, use subtraction. Since the two fractions, $2\frac{3}{4}$ and $\frac{7}{8}$, have different denominators, it is necessary to find a common denominator, which is 8. First, convert $2\frac{3}{4}$ into an improper fraction $\frac{2 \times 4 + 3}{4} = \frac{11}{4}$. Afterwards, convert the fraction, $\frac{11}{4}$, by multiplying both the numerator and denominator by 2: $\frac{11 \times 2}{4 \times 2} = \frac{22}{8}$. Now, both fractions have a common denominator. Thus, subtract the two fractions.

$$\frac{22}{8} - \frac{7}{8} = \frac{15}{8} = 1\frac{7}{8}$$

Therefore, how much water was left in the water cooler is $1\frac{7}{8}$.

8. **(J)**

Joshua wants to buy a bicycle that costs $250 and has a sales tax rate of 8%. This implies that the total amount of money that Joshua needs to pay is 108% of the original cost of the bicycle; 100% accounts for the original cost of the bicycle and 8% accounts for the sales tax. Since $108\% = 1.08$, the total amount needed to pay for the bicycle including sales tax is $\$250 \times 1.08 = \$270$.

9. (D)

The ratio of white paint to green paint in the mixture is $7 : 3$. Set up a proportion to determine the amount of green paint needed when using 28 ounces of white paint.

$$7_{\text{white}} : 3_{\text{green}} = 28_{\text{white}} : x_{\text{green}}$$
$$\frac{7}{3} = \frac{28}{x} \quad \text{Cross multiply}$$
$$7x = 3 \times 28$$
$$x = \frac{3 \times 28}{7}$$
$$x = 12$$

Therefore, 12 ounces of green paint would be mixed with 28 ounces of white paint.

10. (L)

In order to compare numbers, convert the numbers into decimals: $20\% = 0.2$ and $\frac{1}{4} = 0.25$. Afterwards, arrange the decimals from least to greatest.

$$\{0.2, 0.25, 0.8, 1.5\} \quad \Longrightarrow \quad \{20\%, \frac{1}{4}, 0.8, 1.5\}$$

Therefore, (L) is the correct answer.

11. (B)

$\frac{1}{8} = 0.125$. In order to convert a decimal number into a percentage, multiply by 100 and then add a % sign. Therefore, $0.125 \times 100 = 12.5\%$

12. (K)

The baseball team wins 3 out of 4 games. This implies that the team loses 1 out of 4 games. Therefore, the number of games that the team loses out of 60 games is $60 \times \frac{1}{4} = 15$.

13. (D)

17 goes into the number 111 six times; $17 \times 6 = 102$ and $17 \times 7 = 119$. Therefore, the remainder when 111 is divided by 17 is $111 - 102 = 9$.

14. (L)

There are 48 pieces of candy in a box. Joshua eats half of the candies. Thus, there are only $48 \times \frac{1}{2} = 24$ pieces remaining. Since Jason eats $\frac{3}{4}$ of the remaining pieces, there are $\frac{1}{4}$ of the 24 or 6 pieces remaining in the box.

15. (C)

If a $1200 computer is discounted 25% off, that means the new price of the computer is 75% of the original value. Convert 75% into a decimal by moving the decimal point left two places. Therefore, the new price of the computer is $1200 \times 0.75 = 900.

SOLOMON ACADEMY — Distribution or replication of any part of this page is prohibited. — TEST 1 SECTION 2

Answers and Solutions
IAAT Practice Test 1 Section 2

Answers

1. D	2. M	3. C	4. K	5. A
6. L	7. B	8. L	9. B	10. L
11. D	12. J	13. A	14. M	15. C

Solutions

1. (D)

   The pie chart displays the four sports in which athletes participate. The total number of athletes can be determined by adding up the values shown on the pie chart: 350 for golf, 300 for football, 200 for swimming, and 150 for tennis. Therefore, the total number of athletes in Fairfax County is $350 + 300 + 200 + 150 = 1000$.

2. (M)

   1% means 1 out of 100 or $\frac{1}{100}$. There are 150 athletes who participate in tennis out of 1000 total athletes. Therefore, the percentage of athletes playing tennis is equal to $\frac{150}{1000} = \frac{15}{100} = 15\%$.

3. (C)

   According to the pie chart, the most favorite sport is golf with 350 participants. The least favorite sport is tennis with 150 participants. Therefore, there are $350 - 150 = 200$ more athletes that participate in golf than tennis.

4. (K)

   Probability is a measure of how likely an event will happen. The definition of probability of an event, E, is as follows:

   $$\text{Probability (E)} = \frac{\text{The number of favorable outcomes}}{\text{The total number of outcomes}}$$

   Since there are 300 athletes who play football out of 1000 total athletes, the probability that a randomly chosen athlete participates in football is $\frac{300}{1000} = \frac{3}{10}$.

5. (A)

   Observe the graph. The temperature increased gradually throughout from day 1 to 13. Therefore, (A) is the correct answer. Answer choice (B) is incorrect because there was no decrease in temperature. Answer choice (C) is incorrect because the temperature continued to rise after day 9. Lastly, answer choice (D) is incorrect because the temperature rose in between day 1 and day 5.

6. (L)

The temperature did not increase means that there was either a decrease in temperature or no change in temperature. There were three times at which the temperature did not increase: from day 2 to day 3, from day 4 to day 5, and from day 9 to day 10. Therefore, the last time that the temperature did not increase was from day 9 to day 10.

7. (B)

There was a steady increase in temperature between day 5 and day 9. Since the total increase in temperature was $(16-8)°F$ or $8°F$, the increase in temperature per day was $\frac{8°F}{4 \text{ days}} = 2°F$ per day.

8. (L)

There was the largest increase of temperature from day 11 to day 12. Since the temperatures on day 12 and day 11 are $22°F$ and $18°F$, respectively, the largest increase of temperature is equal to $(22-18)°F$ or $4°F$.

9. (B)

Sue went to the store and purchased the listed items with a $20 bill. The total bill came out to be $1.50 + $3.50 + $2.00 + $3.25 = 10.25. Therefore, the amount of change that Sue received was $20.00 - $10.25 = 9.75.

10. (L)

The pattern consists of four elements: ♣, ◇, ♡, ♠. This means that every fourth term, or multiple of four, will have the shape ♠: the 4th term, the 8th term, and the 12th term are ♠. Therefore, the 13th term is ♣, the 14th term is ◇, and finally, the 15th term is ♡.

11. (D)

Add up the number of miles Mr. Rhee ran each day.

$$2.5 + 4 + 3 + 2.4 + 1.3 = 13.2$$

Therefore, the number of miles that Mr. Rhee ran in the five days is 13.2 miles.

12. (J)

Name	Game 1	Game 2	Game 3	Game 4	Total
Alex	1	0	1	3	5
Joshua	0	3	0	5	8
Jason	3	4	4	1	12

The table above has an added column that shows the total score that each person scored on the four games. Therefore, Alex scored the least for the four games.

13. (A)

In order to find Jason's average test score, divide the sum of four game scores by 4.

$$\text{Average} = \frac{\text{Sum of four games}}{4 \text{ games}} = \frac{3+4+4+1}{4}$$
$$= \frac{12}{4}$$
$$= 3$$

Therefore, Jason's average score for the four games is 3.

14. (M)

Name	Game 1	Game 2	Game 3	Game 4
Alex	1	0	1	3
Joshua	0	3	0	5
Jason	3	4	4	1
Total	**4**	**7**	**5**	**9**

The table above has an added row that shows the total score received by the three students on each game. Therefore, the sum of the three scores on game 4 is largest.

15. (C)

In order to find the mean (average) of the temperature, divide the sum of the five temperatures by 5.

$$\text{Mean Temperature} = \frac{\text{Sum of five temperatures}}{5 \text{ days}}$$
$$= \frac{52+52+60+70+76}{5}$$
$$= \frac{310}{5}$$
$$= 62$$

Therefore, the mean of the temperatures is 62.

SOLOMON ACADEMY Distribution or replication of any part of this page is prohibited. TEST 1 SECTION 3

Answers and Solutions
IAAT Practice Test 1 Section 3

Answers

1. C	2. K	3. B	4. L	5. D
6. J	7. B	8. K	9. D	10. M
11. A	12. J	13. C	14. J	15. B

Solutions

1. (C)

    A relation is a set of ordered pairs (x, y). A function is a special type of a relation in which each input, x, is related to exactly one output, y. In answer choice (C), the values of x are repeated and are related to two different values of y: $(3, 2)$ and $(3, -2)$. Thus, the set of ordered pairs in answer choice (C) is not a function. Therefore, (C) is the correct answer.

2. (K)

    Slope describes the steepness and direction of a line. Slope is defined as $\frac{y_2-y_1}{x_2-x_1}$, where (x_1, y_1) and (x_2, y_2) represent two points that a line passes through.

    $$\text{Slope} = \frac{y_2 - y_1}{x_2 - x_1} = \frac{2 - (-2)}{1 - (-1)} = \frac{4}{2} = 2$$

    Therefore, the slope of the line that passes through the points $(-1, -2)$ and $(1, 2)$ is 2.

3. (B)

x	$y = 2x$	(x, y)
1	$y = 2(1) = 2$	$(1, 2)$
2	$y = 2(2) = 4$	$(2, 4)$
3	$y = 2(3) = 6$	$(3, 6)$

    Plug in the x values into the equations listed in answer choices to determine which would yield the desired y values. The equation $y = 2x$ is the only equation that satisfies the three ordered pairs $(1, 2)$, $(2, 4)$, and $(3, 6)$. Therefore, (B) is the correct answer.

4. (L)

    No matter what the values of x are, the values of y are always 3. Therefore, the equation that best represents the three ordered pairs $(1, 3)$, $(2, 3)$, and $(3, 3)$ is $y = 3$.

5. (D)

Plug in $x = \frac{1}{3}$ into the equation $y = \frac{1}{2}(1-x)$ and solve for y.

$$y = \frac{1}{2}\left(1-x\right) = \frac{1}{2}\left(1-\frac{1}{3}\right) = \frac{1}{2}\left(\frac{2}{3}\right) = \frac{1}{3}$$

6. (J)

Substitute 13 for y in the equation $y = 3x + 4$ and solve for x.

$y = 3x + 4$	Substitute 13 for y
$13 = 3x + 4$	Subtract 4 from each side
$3x = 9$	Divide each side by 3
$x = 3$	

Therefore, the value of x is 3.

7. (B)

x	$y = 2x + 1$
2	$y = 2(2) + 1 = 5$
3	$y = 2(3) + 1 = 7$
4	$y = 2(4) + 1 = 9$
5	$y = 2(5) + 1 = 11$

The table above shows a relationship between input, x, and output, y: $y = 2x + 1$. In other words, the value of the output, y, is equal to one more than two times the value of the input, x. Therefore, when the input is 5, the value of the output is $2(5) + 1 = 11$.

8. (K)

Input	Output
1	4
2	7
3	10
4	13
5	16
6	19

Solve this problem by noticing the pattern shown in the table above. As the input value increases by 1, the output value increases by 3. This means that the input value of 4 has an output value of 13, the input value of 5 has an output value of 16, and finally, the input value of 6 has an output value of 19. Therefore, 6 should be in the empty cell.

9. (D)

x	$y = 2x - 2$
1	$y = 2(1) - 2 = 0$
2	$y = 2(2) - 2 = 2$
3	$y = 2(3) - 2 = 4$
4	$y = 2(4) - 2 = 6$

All ordered pairs in each answer choice must satisfy the equation $y = 2x - 2$. The only set of ordered pairs that satisfy the equation is answer choice (D) as shown in the table above. Therefore, (D) is the correct answer.

10. (M)

In order to quickly determine the relationship between the expense and profits of the company, observe the values of the second ordered pair. If the expense is 100, then the profit is 130. This means that the profit is 1.3 times the expense. Therefore, the equation that represents the relationship between expenses and profit of the company is $P = 1.3E$.

11. (A)

x	$y = 3x + 4$
2	$y = 3(2) + 4 = 10$
3	$y = 3(3) + 4 = 13$
4	$y = 3(4) + 4 = 16$
5	$y = 3(5) + 4 = 19$

The number of burgers sold, y, is four more than three times the number of sodas sold, x. This verbal relationship can be expressed as the equation $y = 3x + 4$. All ordered pairs in each table must satisfy this equation to be a solution. Make sure to test out all ordered pairs in a given table and do not assume that the other ordered pairs meet the requirements. The only table that represents $y = 3x + 4$ is answer choice (A) as shown in the table above.

12. (J)

Four times the number of boys, b, can be represented as $4b$. Furthermore, five less than four times the number of boys is $4b - 5$ because 5 is being taken away from four times the number of boys. Thus, the number of girls, g, is five less than four times the number of boys, b, can be represented as $g = 4b - 5$.

13. (C)

The equation of the line is written in slope-intercept form: $y = mx + b$, where m and b represent the slope and y-intercept, respectively. Two lines are considered parallel to each other if they have the same slope but different y-intercepts. Since the slope of the equation $y = -3x + 4$ is -3, the only equation that is parallel to $y = -3x + 4$ is $y = -3x + 8$ in answer choice (C).

14. (J)

In order to find the y-intercept of a line, substitute 0 for x.

$y = 2x + 6$ Substitute 0 for x
$y = 2(0) + 6$ Simplify
$y = 6$

Therefore, the y-intercept of the line $y = 2x + 6$ is 6.

15. (B)

Quantity	10	20	40	80	160	320	640
Unit Price	$10	$9.5	$9	$8.5	$8	$7.5	$7

A t-shirt company provides lower prices if shirts are purchased in larger quantities. As the quantity of shirts is doubled, the unit price of the t-shirt decreases by $0.50. The cost per shirt of an order with 160 shirts is $8, 320 shirts is $7.50, and 640 shirts is $7 as shown in the table above. Therefore, if the quantity reaches 640, the unit price of the t-shirt is $7.

SOLOMON ACADEMY — TEST 1 SECTION 4

Answers and Solutions
IAAT Practice Test 1 Section 4

Answers

1. C	2. L	3. A	4. L	5. B
6. J	7. D	8. K	9. B	10. M
11. D	12. J	13. C	14. K	15. D

Solutions

1. (C)

   Since the rectangles are similar, the ratio of their corresponding sides are equal. Let's set up a proportion in terms of length and width of the rectangles.

   $$\frac{\text{length}}{\text{width}} : \quad \frac{10}{6} = \frac{15}{x}$$

   Therefore, (C) is the correct answer.

2. (L)

   Substitute the values given in the question and evaluate the expression. Use the order of operations: PEMDAS.

   $$\begin{aligned} 2x - y &= 2(3) - (-4) \quad &\text{(Substitute 3 for } x \text{ and } -4 \text{ for } y\text{)} \\ &= 6 + 4 \\ &= 10 \end{aligned}$$

   Therefore, the value of $2x - y$ when $x = 3$ and $y = -4$ is 10.

3. (A)

   In order to solve proportions, cross multiply and solve for x.

   $$\begin{aligned} \frac{3}{8} &= \frac{12}{x} &\text{(Cross multiply)} \\ 3x &= 96 &\text{(Divide each side by 3)} \\ x &= 32 \end{aligned}$$

   Therefore, the value of x is 32.

4. (L)

In order to simplify the expression, use the distributive property: $-a(b-c) = -ab + ac$. Notice that the negative is distributed to both terms inside the parenthesis.

$$-2(3x - 4) = -2(3x) + (-2)(-4) = -6x + 8$$

Therefore, $-2(3x - 4) = -6x + 8$.

5. (B)

Since $x = -6$ and $y = -12$, $\frac{y}{x} = \frac{-12}{-6} = 2$, $xy = (-6)(-12) = 72$, $x + y = -6 + -12 = -18$, and $x - y = -6 - (-12) = 6$. Thus, the expression xy has the largest value. Therefore, (B) is the correct answer.

6. (J)

$$3x + 2 = 17 \quad \text{(Subtract 2 from each side)}$$
$$3x = 15 \quad \text{(Divide each side by 3)}$$
$$x = 5$$

Therefore, the value of x is 5.

7. (D)

Substitute the value given in each answer choice to see if the value satisfies the inequality.

$$11 - x < -4 \quad \text{(Substitute 16 for } x\text{)}$$
$$11 - 16 < -4 \quad \text{(Simplify)}$$
$$-5 < -4 \quad \text{(Inequality holds true)}$$

The only value that satisfies the inequality is 16. Therefore, (D) is the correct answer.

8. (K)

Jason can paint 6 pictures in 3 hours. Set up a proportion in terms of pictures and hours.

$$6_{\text{pictures}} : 3_{\text{hours}} = p_{\text{pictures}} : 7_{\text{hours}}$$
$$\frac{6}{3} = \frac{p}{7}$$

Therefore, (K) is the correct answer.

9. (B)

The volume of a cylinder is defined as $\pi r^2 h$, where r and h is the radius and height respectively. The problem states that the height is 7 and the diameter is 4. Since the radius is half the length of a diameter, $r = 2$. Therefore, the volume of the cylinder is $\pi r^2 h = \pi(2^2)(7) = \pi(4)(7) = 28\pi$.

SOLOMON ACADEMY Distribution or replication of any part of this page is prohibited. TEST 1 SECTION 4

10. (M)

Since $z = 6 + x$, substitute $6 + x$ for z in the equation $x + 2y = z$ and solve for y.

$$\begin{aligned} x + 2y &= z & \text{(Substitute } 6 + x \text{ for } z) \\ x + 2y &= 6 + x & \text{(Subtract } x \text{ from each side)} \\ 2y &= 6 & \text{(Divide each side by 2)} \\ y &= 3 \end{aligned}$$

Therefore, the value of y is 3.

11. (D)

Since $a \blacktriangle b$ is defined as $\dfrac{a+b}{a-b}$,

$$6 \blacktriangle 3 = \frac{6+3}{6-3} = \frac{9}{3} = 3 \tag{1}$$

Therefore, the value of $6 \blacktriangle 3$ is 3.

12. (J)

x is increased by 3, and y is decreased by 8 can be expressed as $x + 3$, and $y - 8$, respectively. Thus,

$$\begin{aligned} x + 3 + y - 8 &= x + y + (3 - 8) & \text{(Since } x + y = z) \\ &= z - 5 \end{aligned}$$

Therefore, if x is increased by 3, and y is decreased by 8, the new value of z is $z - 5$.

13. (C)

Substitute 3 for x and solve for y.

$$\begin{aligned} 3x + 4y &= 25 & \text{(Substitute 3 for } x) \\ 9 + 4y &= 25 & \text{(Subtract 9 from each side)} \\ 4y &= 16 & \text{(Divide each side by 4)} \\ y &= 4 \end{aligned}$$

Therefore, the value of y is 4.

14. (K)

Expression	Verbal Phrase
$4 + x$	Sum of four and a number x
$2(4 + x)$	Two times the sum of four and a number x
$2(4 + x) + 3$	Three more than the sum of four and a number x

Therefore, (K) is the correct answer.

15. (D)

In order to simplify the expression $4x + 3x + 2$, combine like terms. $4x$ and $3x$ are liked terms because each term consists of only a single variable x and the same exponent. Therefore, $4x$ and $3x$ can be combined as $4x + 3x = 7x$. Since 2 is a constant, $7x$ and 2 are unlike terms and cannot be combined. Thus, $4x + 3x + 2 = 7x + 2$. Therefore, (D) is the correct answer.

IAAT PRACTICE TEST 2

SECTION 1
Time — 10 minutes
15 Questions

Directions: Read the information given and choose the best answer for each question. Base your answer only on the information given. The time limit for each section is 10 minutes.

1. $8\frac{1}{4} \div \frac{3}{4} =$

    (A) $\frac{99}{16}$
    (B) $8\frac{1}{3}$
    (C) 11
    (D) 16

2. The temperatures in Minneapolis during four days were 60°F, 65°F, 85°F, and 70°F. What was the average temperature during these four days?

    (J) 68°F
    (K) 70°F
    (L) 72°F
    (M) 74°F

3. Jason earned $560 last summer and spent $\frac{3}{8}$ of his earnings. How much money did he have left?

    (A) $210
    (B) $260
    (C) $350
    (D) $410

4. $15 - 3(7 - 5 + 1) =$

    (J) 3
    (K) 6
    (L) 12
    (M) 36

5. If Q is divisible by 2, 3, and 5, what could be a possible value of Q?

    (A) 65
    (B) 50
    (C) 40
    (D) 30

6. While purchasing a shirt, Joshua paid with a $20 bill. He received 4 dollars and 3 quarters in change. How much was the cost of the shirt?

    (J) $4.75
    (K) $12.75
    (L) $14.25
    (M) $15.25

7. $18 + 4 \times 3 \div (-4) =$

   (A) 15
   (B) 14
   (C) 13
   (D) 12

$-3, 6, -12, 24, \cdots$

8. What is the next number in the sequence above?

   (J) 36
   (K) 48
   (L) -48
   (M) -36

9. What is the value of $3^3 \times 3^{-5}$?

   (A) 3^8
   (B) 9
   (C) 3
   (D) $\frac{1}{9}$

10. What is 12% of 250?

    (J) 20
    (K) 25
    (L) 30
    (M) 35

11. $2.5 \times 1.6 =$

    (A) 3.2
    (B) 4
    (C) 4.6
    (D) 5

12. Which of the following expression has a positive value?

    (J) $5 + (-6)$
    (K) $5 \times (-6)$
    (L) $5 - (-6)$
    (M) $5 \div (-6)$

13. What is the square root of 36?

    (A) 6
    (B) 72
    (C) 360
    (D) 1296

14. The school auditorium can hold 260 people. On the opening night of a heritage night, 214 people attended the show. All but 19 of the people who purchased tickets came. How many tickets were purchased?

   (J) 214
   (K) 233
   (L) 19
   (M) 65

15. Simplify the expression: $\dfrac{5}{16} + \dfrac{7}{10}$

   (A) $\dfrac{82}{80}$
   (B) $\dfrac{41}{40}$
   (C) $1\dfrac{2}{40}$
   (D) $1\dfrac{1}{80}$

STOP

IAAT PRACTICE TEST 2

SECTION 2
Time — 10 minutes
15 Questions

Directions: Read the information given and choose the best answer for each question. Base your answer only on the information given. The time limit for each section is 10 minutes.

Directions: Use the following figure to answer questions 1 – 2.

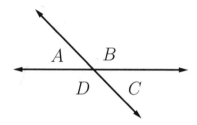

1. If $m\angle A = 62°$, what is the $m\angle B$?

   (A) 28°

   (B) 62°

   (C) 105°

   (D) 118°

2. If $m\angle D = 120°$, what is the $m\angle B$?

   (J) 150°

   (K) 120°

   (L) 60°

   (M) There is not enough information given to answer this question.

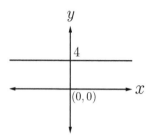

3. What is the value of the slope of the line in the graph above?

   (A) −4

   (B) 0

   (C) 4

   (D) Undefined

4. $\frac{1}{3}$ of the class are boys. If $\frac{1}{4}$ of the boys play piano and $\frac{1}{2}$ of the girls play piano, what fraction of the class plays piano?

   (J) $\frac{1}{12}$

   (K) $\frac{1}{3}$

   (L) $\frac{3}{8}$

   (M) $\frac{5}{12}$

55

5. Jason has 40 ounces of chocolate bars in a jar. If he eats 40% of the chocolate bars, how much chocolate bars, in ounces, are remaining in the jar?

   (A) 20
   (B) 24
   (C) 28
   (D) 32

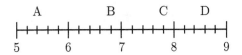

6. If the total weight of eight packages is 62.4 pounds, which of the letters displayed above represent the average weight of a package?

   (J) A
   (K) B
   (L) C
   (M) D

---

7. How many different three-digit numbers are possible if the digits 1, 2, and 3 are used exactly once?

   (A) 5
   (B) 6
   (C) 7
   (D) 8

---

8. Joshua made fruit punch using 1300 mL of apple juice, 2400 mL of grape juice, and 1700 mL of strawberry juice. After making the fruit punch, Joshua equally poured the fruit punch into 6 pitchers. At his party, the attending guest drank 2 pitchers. How much fruit punch did Joshua have left over after the party?

   (J) 900 mL
   (K) 1800 mL
   (L) 3600 mL
   (M) 4800 mL

Day	Miles
Monday	7.25
Tuesday	15
Wednesday	10.5
Thursday	7

9. Jason runs Monday through Friday. The table above shows the number of miles Jason ran from Monday to Thursday. If Jason wants to run a total of 50 miles, how many more miles does he need to run on Friday?

   (A) 9.25
   (B) 10.25
   (C) 11.25
   (D) 12.25

$$\Gamma, \Delta, \Sigma, \Psi, \Omega, \Gamma, \Delta, \cdots$$

10. What is the Greek letter that represents the 23rd term of the pattern shown above?

    (J) Σ

    (K) Δ

    (L) Γ

    (M) Ψ

M	T	W	R	F	Total
35	14	3	17	22	

11. The table above shows the amount of cookies sold Monday through Friday. What is the total number of cookies sold Monday through Friday?

    (A) 89

    (B) 91

    (C) 95

    (D) 99

Directions: Use the following table to answer questions 12 – 14.

Name	Game 1	Game 2	Game 3	Game 4
Alex	81	89	95	95
Joshua	82	75	90	73
Jason	100	63	92	95

12. What is the total points that Joshua scored on the four games?

    (J) 350

    (K) 340

    (L) 330

    (M) 320

13. What is the closest to the mean (average) points in game 3?

    (A) 92

    (B) 90

    (C) 85

    (D) 83

14. How many more points did Jason score than Joshua?

    (J) 24

    (K) 27

    (L) 30

    (M) 33

$$\{-2,\ 6,\ 9,\ -5,\ 2\}$$

15. What is the median of the set above?

    (A) -5

    (B) -2

    (C) 2

    (D) 9

IAAT PRACTICE TEST 2

SECTION 3
Time — 10 minutes
15 Questions

Directions: Read the information given and choose the best answer for each question. Base your answer only on the information given. The time limit for each section is 10 minutes.

1. If $y = 2x - 5$ and $y = 5$, what is the value of x?

   (A) 6
   (B) 5
   (C) 4
   (D) 3

2. Which of the following line has positive slope?

   (J) $x = -2$
   (K) $y = 2$
   (L) $y = 2x - 1$
   (M) $y = -2x + 1$

3. Sue has S number of marbles, which is one less than three times the number of marbles that Jason has, J. Which of the following represents this situation?

   (A) $J = S - 3$
   (B) $J = 3S - 1$
   (C) $S = J - 3$
   (D) $S = 3J - 1$

4. Which of the following table does **NOT** represent a function?

   (J)
x	1	1	1	1
y	-2	-1	0	1

   (K)
x	-2	-1	0	1
y	3	3	3	3

   (L)
x	1	2	3	4
y	2	3	4	5

   (M)
x	0	5	2	1
y	2	3	3	5

5. Every dollar that Joshua earns, Mr. Rhee earns three. If Mr. Rhee earned $18, how much money did Joshua earn?

   (A) $6
   (B) $15
   (C) $21
   (D) $54

6. When bowling, Sue always scores 25 more than twice of Joshua's score. Which of the following tables represent this relationship?

(J)
Sue	25	30	35
Joshua	75	85	95

(K)
Sue	75	85	95
Joshua	25	30	35

(L)
Sue	75	135	175
Joshua	50	80	105

(M)
Sue	50	80	105
Joshua	75	135	175

Input	Output
2	4
3	9
4	16
5	

7. Observe the numbers in the two columns in order to determine which of the following value should be in the empty cell.

(A) 22
(B) 23
(C) 24
(D) 25

x	y
2	1
3	1.5
4	2
5	2.5

8. The table shows four pairs of x and y values. Which is true for all values in the table shown above?

(J) $y = 2x$
(K) $y = \frac{1}{2}x$
(L) $y = x^2$
(M) $y = \sqrt{x}$

9. When $y = -2$, what is the value of x if $y = 3 - x$?

(A) -5
(B) 3
(C) 4
(D) 5

10. What is the slope of the line that passes through two points $(2, 4)$ and $(-3, 2)$?

(J) $-\frac{5}{2}$
(K) $-\frac{2}{5}$
(L) $\frac{5}{2}$
(M) $\frac{2}{5}$

11. Which of the following table best represents the following verbal relationship? The number of cups of coffee sold, y, is 3 less than two times the number of cookies sold, x.

(A)
x	2	3	4	5
y	1	3	5	7

(B)
x	1	2	3	4
y	3	5	7	9

(C)
x	0	1	2	3
y	1	2	3	4

(D)
x	1	2	3	4
y	2	2.5	3	3.5

12. $y = 4x + 15$ represents a straight line on the xy-plane. What is the y-intercept of the line?

(J) -15

(K) -4

(L) 4

(M) 15

13. Which of the following ordered pair does the line $y = x + 1$ passes through?

(A) $(0, 0)$

(B) $(0, 1)$

(C) $(1, 0)$

(D) $(1, 1)$

14. Which of the following equations best represent the following verbal relationship? The number of green marbles, x, is six less than five times the number of blue marbles, y.

(J) $x = 5y - 6$

(K) $x = 6y - 5$

(L) $y = 5x + 6$

(M) $y = 6x + 5$

x	y
1	3
2	9
3	27
4	

15. Observe the numbers in the two columns in order to determine which of the following value should be in the empty cell.

(A) 36

(B) 54

(C) 81

(D) 96

STOP

# IAAT PRACTICE TEST 2

SECTION 4
Time — 10 minutes
15 Questions

Directions: Read the information given and choose the best answer for each question. Base your answer only on the information given. The time limit for each section is 10 minutes.

1. Which of the following verbal phrase is represented by $\dfrac{x}{3}$?

    (A) The quotient of x and 3.

    (B) The quotient of 3 and x.

    (C) 3 divided by x.

    (D) The product of x and 3.

$$-13 \square 4 = -9$$

2. Which of the following can be placed in the box to make the equation true?

    (J) +

    (K) −

    (L) ×

    (M) ÷

---

3. Which of the following inequality is true?

    (A) $-3 > 1$

    (B) $-3 > 0$

    (C) $-3 > -4$

    (D) $-3 < -4$

$$x + 4 > 4$$

4. Which of the following value of x satisfies the inequality above?

    (J) −2

    (K) −1

    (L) 0

    (M) 1

---

5. Simplify the expression: $3x + 4 - x + 3$

    (A) $2x + 1$

    (B) $2x + 7$

    (C) $4x + 1$

    (D) $4x + 7$

---

6. If $x = -6$ and $y = 7$, what is the value of $y - x$?

    (J) 13

    (K) 1

    (L) −1

    (M) −13

25 percent of a number, x

7. Which of the following expression represents the verbal phrase above?

   (A) $\frac{x}{25}$
   (B) $\frac{x}{0.25}$
   (C) $25x$
   (D) $0.25x$

Eight more than the square of a number, z.

8. How is the following verbal phrase above expressed algebraically?

   (J) $8 + 2z$
   (K) $8z + 2$
   (L) $z^2 + 8$
   (M) $(z + 8)^2$

9. Solve for x: $6x - 4 = 44$

   (A) 7
   (B) 8
   (C) 9
   (D) 10

10. If $\frac{6}{x} = 3$, what is the value of x?

    (J) 4
    (K) 3
    (L) 2
    (M) 1

Three less than the product of x and 2 is 13

11. Solve for x of the following verbal phrase above.

    (A) 8
    (B) 6
    (C) 4
    (D) 2

12. Which of the following is the solution to the inequality $-4x \leq 12$?

    (J) $x \geq -3$
    (K) $x \leq -3$
    (L) $x \geq 3$
    (M) $x \leq 3$

13. The formula for the volume of a rectangular box is $\ell \times w \times h$, where ℓ, w, and h represent the length, width, and height, respectively. What is the volume of a rectangular box with a length of 4, width of 3, and height of 5?

    (A) 12
    (B) 36
    (C) 48
    (D) 60

14. Jason puts 3 cupcakes into each of 12 boxes. Let c represent the total number of cupcakes he had. Which of the following best represents c ?

   (J) $c = 12 + 3$

   (K) $c = 12 - 3$

   (L) $c = 12 \div 3$

   (M) $c = 12 \times 3$

15. Mr. Rhee had $500 in his savings account. If he saves $300 per week, in how many weeks will he have saved a total of $3500 in his savings account?

   (A) 11

   (B) 10

   (C) 9

   (D) 8

STOP

SOLOMON ACADEMY — Distribution or replication of any part of this page is prohibited. — TEST 2 SECTION 1

Answers and Solutions
IAAT Practice Test 2 Section 1

Answers

1. C	2. K	3. C	4. K	5. D
6. M	7. A	8. L	9. D	10. L
11. B	12. L	13. A	14. K	15. D

Solutions

1. (C)

   Convert any mixed numbers into improper fractions. Thus, $8\frac{1}{4}$ can be written as $\frac{8\cdot 4+1}{4} = \frac{33}{4}$.

   $$\frac{33}{4} \div \frac{3}{4} = \frac{33}{4} \times \frac{4}{3} = \frac{33}{3} = 11$$

2. (K)

   In order to find the average temperature, divide the sum of four temperatures by 4.

   $$\text{Average Temperature} = \frac{60+65+85+70}{4} = \frac{280}{4} = 70$$

   Therefore, the average temperature during these four days is $70°F$.

3. (C)

   Jason earned $560 and spent $\frac{3}{8}$ of his earnings. This means that the amount of money he has left is equivalent to $\frac{5}{8}$ of $560. Therefore, the amount of money Jason has left is $\frac{5}{8} \times \$560$ or $350.

4. (K)

   When evaluating numerical expressions, use the order of operations: PEMDAS. This means to evaluate expressions in the parenthesis first, then exponents, then multiplication and division from left to right, and finally addition and subtraction from left to right.
   Therefore, $15 - 3(7 - 5 + 1) = 15 - 3(3) = 6$.

5. (D)

   The least common multiple of 2, 3, and 5 is 30. Thus, the possible values of Q are multiples of 30: $30, 60, 90, \cdots$. Therefore, (D) is the correct answer.

6. (M)

   Subtract $20 by the total amount of change received in order to determine the cost of the shirt. 4 dollars and 3 quarters are equivalent to $4 + 0.75 = 4.75$. Thus, Joshua received $4.75 in change. Therefore, the cost of the shirt is $\$20 - \$4.75 = \$15.25$.

7. (A)

When evaluating numerical expressions, use the order of operations: PEMDAS. This means to evaluate expressions in the parenthesis first, then exponents, then multiplication and division from left to right, and finally addition and subtraction from left to right. Note when multiplying or dividing by a negative number, the result is negative.

$$18 + 4 \times 3 \div (-4) = 18 + 12 \div (-4)$$
$$= 18 + (-3)$$
$$= 15$$

8. (L)

$$-3,\ 6,\ -12,\ 24,\ \cdots$$

The sequence above suggests a pattern such that multiply preceding term by -2 to get the next term. For example, $-3 \times -2 = 6$, and $6 \times -2 = -12$. Therefore, the next term after 24 in the sequence is 24×-2 or -48.

9. (D)

Use the properties of exponents: $a^x \times a^y = a^{x+y}$ and $a^{-x} = \frac{1}{a^x}$.

$$3^3 \times 3^{-5} = 3^{3+(-5)} = 3^{-2}$$
$$= \frac{1}{3^2} = \frac{1}{9}$$

Therefore, the value of $3^3 \times 3^{-5}$ is $\frac{1}{9}$.

10. (L)

Convert 12% into a decimal by moving the decimal point two places to the left: 12% = 0.12. In order to determine 12% of 250, find the product of 0.12 and 250. To multiply numbers containing decimal points, simply multiply ignoring the decimal points first. Afterwards, place the decimal point into the solution equivalent to the amount of decimal places the two numbers have together. For example, change 250×0.12 into $250 \times 12 = 3000$. Since 250 has no decimal places and 0.12 has two numbers after the decimal point, the solution will contain two decimal places. Therefore, $250 \times 0.12 = 30$.

11. (B)

To multiply decimal numbers, simply multiply ignoring the decimals points first. Afterwards, place the decimal point into the solution equivalent to the amount of decimal places the two numbers have together. Therefore, change 2.5×1.6 into $25 \times 16 = 400$. Since 2.5 has one number after the decimal point and 1.6 has one number after the decimal point, the solution will contain two decimal places. Therefore, $2.5 \times 1.6 = 4$.

12. (L)

Expression	Value
$5 + (-6)$	-1
$5 \times (-6)$	-30
$5 - (-6)$	11
$5 \div (-6)$	$-\frac{5}{6}$

The only expression that has a positive value is $5 - (-6) = 11$ as shown in the table above. Therefore, (L) is the correct answer.

13. (A)

In order to square a number, multiply it by itself. A square root of a number is the opposite in that it is a value that can be multiplied by itself to obtain the original number. Therefore, the square root of 36 is 6 because $6 \times 6 = 36$. Therefore, (A) is the correct answer.

14. (K)

The maximum capacity of the auditorium is information that is not necessary to solve the problem. 214 people attended the show and all but 19 of the people who purchased came. In other words, 19 people who purchased tickets were absent. Therefore, the total number of tickets purchased is $214 + 19 = 233$.

15. (D)

When adding fractions with uncommon denominators, it is necessary to determine the least common multiple (LCM) of the uncommon denominators. The least common multiple of 16 and 10 is 80. Convert the first fraction by multiplying both the numerator and denominator by 5: $\frac{5 \times 5}{16 \times 5} = \frac{25}{80}$. Likewise, convert the second fraction by multiplying both the numerator and denominator by 8: $\frac{7 \times 8}{10 \times 8} = \frac{56}{80}$. Since both fractions have a common denominator, add two fractions.

$$\frac{5}{16} + \frac{7}{10} = \frac{25}{80} + \frac{56}{80} = \frac{81}{80} = 1\frac{1}{80}$$

Therefore, $\frac{5}{16} + \frac{7}{10} = 1\frac{1}{80}$.

SOLOMON ACADEMY — TEST 2 SECTION 2

Answers and Solutions
IAAT Practice Test 2 Section 2

Answers

1. D	2. K	3. B	4. M	5. B
6. L	7. B	8. L	9. B	10. J
11. B	12. M	13. A	14. L	15. C

Solutions

1. (D)

    Angle A and angle B are supplementary angles. In order for two angles to be considered supplementary, the sum of the measures of angles must add up to 180 degrees. A straight line has 180 degrees. Since the measure of angle A is $62°$, the measure of angle B is $180 - 62 = 118°$.

2. (K)

    When two lines intersect each other, two pairs of vertical angles are formed. Angle B and angle D are vertical angles and congruent. Since the measure of angle D is $120°$, the measure of angle B is also $120°$.

3. (B)

    The slope represents the steepness and direction of a line and is defined as $\frac{\text{rise}}{\text{run}}$ or $\frac{\text{vertical change}}{\text{horizontal change}}$. Note that the slope of a horizontal line is 0 and the slope of a vertical line is undefined. Since the line shown in the graph is horizontal, the slope of the line is 0.

4. (M)

    $\frac{1}{3}$ of the class are boys. This means that $1 - \frac{1}{3}$ or $\frac{2}{3}$ of the class are girls. Since $\frac{1}{4}$ of the boys play piano, $\frac{1}{3} \times \frac{1}{4} = \frac{1}{12}$ represents the fraction of boys who play piano out of the class. Since $\frac{1}{2}$ of the girls play piano, $\frac{2}{3} \times \frac{1}{2} = \frac{1}{3}$ represents the fraction of girls who play piano out of the class. Thus, add the two fractions to determine what fraction of the class plays piano.

    $$\frac{1}{12} + \frac{1}{3} = \frac{1}{12} + \frac{4}{12} = \frac{5}{12}$$

    Therefore, $\frac{5}{12}$ of the class plays piano.

5. (B)

    If Jason eats 40% of the chocolate bars, it means that $100 - 40 = 60\%$ of the chocolate bars are remaining. Therefore, $0.6 \times 40 = 24$ ounces of the chocolate bars are remaining in the jar.

6. (L)

The average, or mean, is defined as the sum of all elements divided by the number of elements. Since there are 8 packages that weigh a total of 62.4 pounds, the average weight of a package is $\frac{62.4}{8} = 7.8$ pounds which is depicted by the letter C on the graph.

7. (B)

The number of three-digit numbers that can be created using the digits 1, 2, and 3 exactly once is 6: 123, 132, 213, 231, 312, and 321. Therefore, (B) is the correct answer.

8. (L)

Joshua made fruit punch using 1300 mL of apple juice, 2400 mL of grape juice, and 1700 mL of strawberry juice. Thus, the total amount of the fruit punch made is $1300 + 2400 + 1700 = 5400$ mL. Since the total amount of fruit punch were poured into 6 pitchers, each pitcher contained $\frac{5400}{6} = 900$ mL. At the party, the guests drank 2 out of the 6 pitches. Thus, 4 pitchers were remaining. Therefore, the amount of fruit punch that Joshua have left over was $4 \times 900 = 3600$ mL.

9. (B)

Jason wants to run a total of 50 miles from Monday to Friday. During Monday through Thursday, Jason ran $7.25 + 15 + 10.5 + 7 = 39.75$ miles. Therefore, Jason must run $50 - 39.75 = 10.25$ miles on Friday.

10. (J)

The pattern consists of five Greek letters: Γ, Δ, Σ, Ψ, Ω. This means that every fifth term, or multiple of five, will have the Greek letter Omega, or Ω. Therefore, the 5th term is Ω, the 10th term is Ω, the 15th term is Ω, and the 20th term is Ω. The 21st term is Γ, 22nd term is Δ, and thus the 23rd term is Σ.

11. (B)

Add up the number of cookies sold each day. The total number of cookies sold in the five days is $35 + 14 + 3 + 17 + 22 = 91$ cookies.

12. (M)

Name	Game 1	Game 2	Game 3	Game 4	Total
Joshua	82	75	90	73	320

The table above shows the total points that Joshua scored on the four games. Therefore, Joshua scored 320 points on the four games.

13. (A)

In order to find the mean points of game 3, divide the total number of points by the total number of participants.

$$\text{Average} = \frac{\text{Total Number of Points}}{\text{Number of Participants}} = \frac{95 + 90 + 92}{3} = \frac{277}{3} = 92\frac{1}{3}$$

Thus, the closest to the mean points in Game 3 is 92. Therefore, (A) is the correct answer.

14. (L)

Name	Game 1	Game 2	Game 3	Game 4	Total
Joshua	82	75	90	73	**320**
Jason	100	63	92	95	**350**

Since Jason scored 350 points and Joshua scored 320 points, Jason scored 30 points more than Joshua did.

15. (C)

$$\{-2, 6, 9, -5, 2\} \implies \{-5, -2, 2, 6, 9\}$$

In order to find the median of the set, rearrange the numbers in the set from least to greatest as shown above and find the middle number. Therefore, the median of the set is 2.

SOLOMON ACADEMY   TEST 2 SECTION 3

Answers and Solutions
IAAT Practice Test 2 Section 3

Answers

1. B	2. L	3. D	4. J	5. A
6. K	7. D	8. K	9. D	10. M
11. A	12. M	13. B	14. J	15. C

Solutions

1. (B)

   Substitute 5 for y in the equation, $y = 2x - 5$, and solve for x.

   $y = 2x - 5$ (Substitute 5 for y)
   $5 = 2x - 5$ (Add 5 to each side)
   $2x = 10$
   $x = 5$

   Therefore, the value of x is 5.

2. (L)

   $x = -2$ is a vertical line whose slope is undefined. $y = 2$ is a horizontal line whose slope is zero. $y = 2x - 1$ is written in slope-intercept form, where the slope is 2 and the y-intercept is -1. Therefore, $y = 2x - 1$ has positive slope.

3. (D)

Verbal Phrase	Expression
Three times the number of marbles that Jason has, J	$3J$
One less than three times the number of marbles that Jason has, J	$3J - 1$

   Thus, the equation that represents the verbal phrase is $S = 3J - 1$. Therefore, (D) is the correct answer.

4. (J)

   A function relates an input to an output so that an input x cannot have more than one value for an output y. In answer choice (J), the x value, 1, repeats and has four different y values: $(1, -2)$, $(1, -1)$, $(1, 0)$, and $(1, 1)$. Therefore, by definition, the table in answer choice (J) does NOT represent a function.

SOLOMON ACADEMY — TEST 2 SECTION 3

5. (A)

   Every dollar that Joshua earns, Mr. Rhee earns three. Since Mr. Rhee earned $18, Joshua earned one-third that amount. Therefore, Joshua earned $\frac{18}{3} = \$6$.

6. (K)

   When bowling, Sue, y, always scores 25 more than twice of Joshua's score, x.

Sue (y)	75	85	95
Joshua (x)	25	30	35

   This verbal relationship can be written as $y = 2x + 25$ which is depicted by the table in answer choice (K).

7. (D)

Input (x)	Output ($y = x^2$)
2	$y = 2^2 = 4$
3	$y = 3^2 = 9$
4	$y = 4^2 = 16$
5	$y = 5^2 = 25$

   The table above shows a relationship that can be expressed as $y = x^2$. In other words, in order to obtain the value of the output y, square the value of the input x. Therefore, when the input is 5, the value of the output is $5^2 = 25$.

8. (K)

   The value of output, y is always half the value of the input, x. The table shows a relationship that can be expressed as $y = \frac{1}{2}x$. Therefore, (K) is the correct answer.

9. (D)

   Substitute -2 for y in the equation $y = 3 - x$ and solve for x.

   $$y = 3 - x \quad \text{(Substitute } -2 \text{ for } y\text{)}$$
   $$-2 = 3 - x \quad \text{(Subtract 3 from each side)}$$
   $$-5 = -x \quad \text{(Multiply each side by } -1\text{)}$$
   $$x = 5$$

   Therefore, the value of x is 5.

10. (M)

Slope describes the steepness and direction of the line and is defined as $\frac{y_2-y_1}{x_2-x_1}$. (x_1, y_1) and (x_2, y_2) are represented by $(2, 4)$ and $(-3, 2)$ respectively. Thus,

$$\text{Slope} = \frac{y_2 - y_1}{x_2 - x_1} = \frac{2 - 4}{(-3) - 2} = \frac{-2}{-5} = \frac{2}{5}$$

Therefore, the slope of the line that passes through the points $(2, 4)$ and $(-3, 2)$ is $\frac{2}{5}$.

11. (A)

x	$y = 2x - 3$
2	$y = 2(2) - 3 = 1$
3	$y = 2(3) - 3 = 3$
4	$y = 2(4) - 3 = 5$
5	$y = 2(5) - 3 = 7$

The number of cups of coffee sold, y, is 3 less than two times the number of cookies sold, x. This verbal relationship can be expressed as the equation $y = 2x - 3$. All ordered pairs in each table must satisfy this equation to be a solution. Make sure to test out all ordered pairs in a given table and do not assume that the other ordered pairs satisfy the equation. The only table that represents $y = 2x - 3$ is answer choice (A) as shown in the table above.

12. (M)

The equation $y = 4x + 15$ represents a line on the xy-plane. The slope of the line is 4 and the y-intercept of the line is 15. Therefore, (M) is the correct answer.

13. (B)

Substitute x-value and y-value into the equation to find out which ordered pair the line $y = x + 1$ passes through. As shown below, the only ordered pair that satisfies the equation is $(0, 1)$.

$y = x + 1$ \qquad (Substitute 0 for x and 1 for y)

$1 = 0 + 1$ \qquad (Equality holds true)

Therefore, (B) is the correct answer.

14. (J)

Verbal Phrase	Expression
Five times the number of blue marbles, y	$5y$
Six less than five times the number of blue marbles, y	$5y - 6$
The number of green marbles, x, is six less than five times the number of blue marbles, y	$x = 5y - 6$

Therefore, (J) is the correct answer.

15. (C)

Observe the values in the table below to determine the pattern. As the value of x increases by 1, the value of y is multiplied by 3.

x	y
1	3
2	9
3	27
4	81

Thus, when $x = 4$, the y-value is $27 \times 3 = 81$. Therefore, (C) is the correct answer.

SOLOMON ACADEMY — TEST 2 SECTION 4

Answers and Solutions
IAAT Practice Test 2 Section 4

Answers

1. A	2. J	3. C	4. M	5. B
6. J	7. D	8. L	9. B	10. L
11. A	12. J	13. D	14. M	15. B

Solutions

1. (A)

   $\dfrac{x}{3}$ is verbally stated in two ways: the quotient of x and 3 or x divided by 3. Therefore, (A) is the correct answer.

2. (J)

   In order to make the equation true, the addition sign must be placed into the box.

   $$-13 + 4 = -9$$

   Therefore, (J) is the correct answer.

3. (C)

   The only true inequality in the answer choices is $-3 > -4$. Therefore, (C) is the correct answer.

4. (M)

   $$x + 4 > 4 \qquad \text{(Subtract 4 from each side)}$$
   $$x > 0$$

   Since $x > 0$, x must be greater than 0. Therefore, (M) is the correct answer.

5. (B)

   In order to simplify the expression $3x + 4 - x + 3$, combine like terms. $3x$ and $-x$ are like terms because each term consists of x and has the same exponent. Thus, $3x$ and $-x$ can be combined as $3x - x = 2x$. Since 4 and 3 are both constants, they can be simplified as $4 + 3 = 7$. $2x$ and 7 are unlike terms and cannot be combined. Therefore, $3x + 4 - x + 3 = 2x + 7$.

6. (J)

When substituting a negative number, it is recommended to use a parenthesis in order to avoid confusion. Subtracting a negative number means addition.

$$y - x = 7 - (-6) \quad \text{(Substitute 7 for } y \text{ and } -6 \text{ for } x\text{)}$$
$$= 7 + 6 \quad \text{(Subtracting a negative number means addition)}$$
$$= 13$$

Therefore, the value of $y - x$ when $x = -6$ and $y = 7$ is 13.

7. (D)

One percent means one out of 100. Thus, 25 percent means 25 out of 100 or $\frac{25}{100} = 0.25$. Therefore, 25 percent of a number, x, can be expressed as $0.25x$.

8. (L)

The square of a number means that the number is multiplied by itself. For example, the square of 2 means $2^2 = 2 \times 2 = 4$. Thus, the square of a number, z, can be expressed as $z \times z = z^2$. Therefore, eight more than the square of a number, z, can be expressed as $z^2 + 8$.

9. (B)

In order to solve for x in an equation, use the reverse order of operations SADMEP and inverse operations.

$$6x - 4 = 44 \quad \text{(Add 4 to each side)}$$
$$6x = 48 \quad \text{(Divide each side by 6)}$$
$$x = 8$$

Therefore, the solution to $6x - 4 = 44$ is $x = 8$.

10. (L)

Since $\dfrac{6}{x} = \dfrac{3}{1}$, cross multiply and solve for x.

$$\frac{6}{x} = \frac{3}{1} \quad \text{(Cross multiply)}$$
$$3x = 6 \quad \text{(Divide each side by 3)}$$
$$x = 2$$

Therefore, the solution to $\dfrac{6}{x} = 3$ is $x = 2$.

11. (A)

Translate the verbal phrase into a mathematic equation and then solve for x. In order to solve for x in an equation, use the reverse order of operations SADMEP and inverse operations. 3 less than the product of x and 2 is 13 can be expressed as $2x - 3 = 13$.

$$2x - 3 = 13 \qquad \text{(Add 3 to each side)}$$
$$2x = 16 \qquad \text{(Divide each side by 2)}$$
$$x = 8$$

Therefore, the solution to $2x - 3 = 13$ is $x = 8$.

12. (J)

Note that when each side of the inequality is multiplied or divided by a negative number, the inequality symbol must be reversed.

$$-4x < 12 \qquad \text{(Divide each side by } -4)$$
$$x > -3 \qquad \text{(Reverse the inequality symbol)}$$

Therefore, the solution to $-4x < 12$ is $x > -3$.

13. (D)

The volume of a rectangular box is $\ell \times w \times h$, where ℓ, w, and h represent the length, width, and height, respectively. Since the length is 4, the width is 3, and the height is 5, the volume of the rectangular box is $4 \times 3 \times 5 = 60$.

14. (M)

Jason puts 3 cupcakes into each of 12 boxes. In other words, each of the 12 boxes contain 3 cupcakes. Therefore, the total number of cupcakes Jason had is $c = 12 \times 3$.

15. (B)

Mr. Rhee had $500 in his savings account and wants to save the total amount of $3500. That implies that he needs to save $3500 - $500 = $3000 additionally. Since Mr. Rhee saves $300 per week, the number of weeks it takes to save $3000 is $\frac{\$3000}{\$300} = 10$ weeks. Therefore, Mr. Rhee needs 10 weeks to save the total amount of $3500.

IAAT PRACTICE TEST 3

SECTION 1
Time — 10 minutes
15 Questions

Directions: Read the information given and choose the best answer for each question. Base your answer only on the information given. The time limit for each section is 10 minutes.

1. Evaluate: $3630 \div 30$

   (A) 0.121
   (B) 1.21
   (C) 12.1
   (D) 121

2. Snowboard Waxing Company charges $7 every 100 square inches to wax a snowboard. If you have four boards with an area of 300 square inches each, how much will it cost to have all snowboards waxed?

   (J) $72
   (K) $84
   (L) $99
   (M) $105

3. Evaluate: $49 \div \frac{7}{9}$

   (A) 63
   (B) 42
   (C) 38
   (D) 35

4. A tennis player wins on the average 7 out of 12 matches. If the tennis player played 72 matches, how many matches did he lose?

   (J) 28
   (K) 30
   (L) 42
   (M) 49

5. Evaluate: $151.5 \div 1.5$

   (A) 1010
   (B) 101
   (C) 10.1
   (D) 1.01

6. Venus is 40 million kilometers away from Earth. What is 40 million written in scientific notation?

   (J) 4×10^{-7}
   (K) 4×10^{-6}
   (L) 4×10^{7}
   (M) 4×10^{8}

7. How many hours and minutes are in between 7:40 pm and 5:10 am of the next morning?

   (A) 7 Hours and 30 Minutes
   (B) 8 Hours and 30 Minutes
   (C) 9 Hours and 30 Minutes
   (D) 10 Hours and 30 Minutes

8. $2.1 + 0.3001 + 0.089 =$

   (J) 2.4114
   (K) 2.49
   (L) 2.4891
   (M) 5.1901

9. What is the greatest common factor (GCF) of 56 and 84?

   (A) 2
   (B) 4
   (C) 28
   (D) 42

10. A rectangle has a width of 7 inches. The length is 50% more than the width. Find the perimeter of the rectangle.

    (J) 21
    (K) 28
    (L) 35
    (M) 42

11. You have 5 types of bread, 3 types of meat, and 2 types of cheese. How many different sandwiches can you make if each sandwich has only 1 type of bread, 1 type of meat, and 1 type of cheese?

    (A) 30
    (B) 25
    (C) 20
    (D) 10

12. Evaluate: 0.56×0.23

    (J) 0.01288
    (K) 0.1288
    (L) 1.288
    (M) 12.88

13. The price of a soda is 60% of the price of a burger. Mr. Rhee paid $8 for a sodas and a burgers. What is the price of the soda?

    (A) $3
    (B) $4
    (C) $5
    (D) $9

14. Jason reads a book that contains 45 pages. How many times does the digit "4" appear on the page numbers that Jason read?

   (J) 8
   (K) 9
   (L) 10
   (M) 11

15. Evaluate: $(-2)^2 - 4(-3)$

   (A) 0
   (B) 6
   (C) 12
   (D) 16

STOP

SOLOMON ACADEMY   Distribution or replication of any   TEST 3 SECTION 2
                         part of this page is prohibited.

IAAT PRACTICE TEST 3

SECTION 2
Time — 10 minutes
15 Questions

Directions: Read the information given and choose the best answer for each question. Base your answer only on the information given. The time limit for each section is 10 minutes.

Directions: Use the following graph to answer questions 1 – 4.

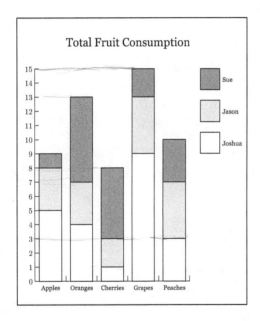

1. What is the mean amount of grapes consumed by three people?

    (A) 5
    (B) 6
    (C) 8
    (D) 9

2. Which fruit did Sue consume the most?

    (J) Peaches
    (K) Grapes
    (L) Oranges
    (M) Cherries

3. Which fruit did Jason consume the least?

    (A) Cherries
    (B) Oranges
    (C) Peaches
    (D) Grapes

4. What is the total number of fruits consumed by Joshua?

    (J) 18
    (K) 20
    (L) 22
    (M) 24

Directions: Use the following graph to answer questions 5 – 8.

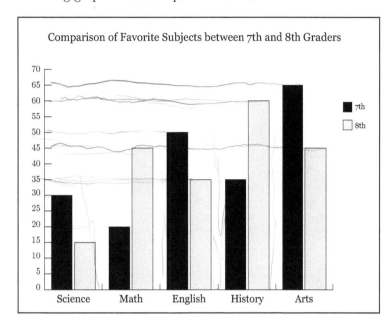

5. How many people were surveyed in all?

   (A) 300
   (B) 400
   (C) 450
   (D) 500

6. How many more 7th graders than 8th graders preferred English as their favorite subject?

   (J) 15
   (K) 25
   (L) 35
   (M) 50

7. What percent of students preferred arts as their favorite subject?

   (A) 12.5%
   (B) 22.5%
   (C) 27.5%
   (D) 32.5%

8. How many more students preferred history over English as their favorite subject?

   (J) 25
   (K) 15
   (L) 12
   (M) 10

Favorite Color	Students
Red	9
Blue	17
Green	8
White	13
Other	3

9. Fifty students were asked to name their one favorite color. The chart above shows the results. What percent of students named blue?

   (A) 34%
   (B) 17%
   (C) 12%
   (D) 8.5%

10. Joshua, Jason, and 13 other students line up in a row to get lunch. Joshua is the 7th from the front and Jason is 3rd from the back. How many students are there between Joshua and Jason? (Do not count Joshua and Jason.)

    (J) 4
    (K) 5
    (L) 6
    (M) 7

M	T	W	R	F
22	52		17	33

11. The table above shows the amount of books sold in the week. If the total number of books sold is 189, which of the following value belongs in the empty cell?

    (A) 60
    (B) 65
    (C) 68
    (D) 75

12. Four soccer teams A, B, C, and D compete against each other in a tournament. Each team plays against every other team only once. How many games are played in all?

    (J) 8
    (K) 7
    (L) 6
    (M) 5

13. 2 pears and 2 apples collectively weigh 24 ounces. 3 pears and 2 apples collectively weigh 29 ounces. Assuming that all the pears and apples respectively weigh the same, how much does a pear weigh?

    (A) 2
    (B) 3
    (C) 4
    (D) 5

Extra-Curricular	Percent
Outdoor Activities	26
Music	21
Chess	7
Reading	18
Playing Games	13
Others	15

14. Jason surveyed students at his school and asked each to select one favorite extra-curricular activity. The percents of the total number of students who responded for each activity are displayed in the table above. If 54 students selected reading as their favorite extra-curricular activity, how many total number of students responded to Jason's survey?

   (J) 200
   (K) 225
   (L) 250
   (M) 300

15. Mr. Rhee runs 90 miles in two weeks. What is the average number of miles Mr. Rhee run per day if he runs 5 days a week?

   (A) 9
   (B) 15
   (C) 18
   (D) 45

STOP

IAAT PRACTICE TEST 3

SECTION 3
Time — 10 minutes
15 Questions

Directions: Read the information given and choose the best answer for each question. Base your answer only on the information given. The time limit for each section is 10 minutes.

1. What type of relationship is shown in the scatter plot below?

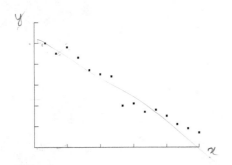

   (A) No Relationship
   (B) Positive Relationship
   (C) Negative Relationship
   (D) Not enough information given

2. Which equation represents the following situation? In order to make a reservation for a birthday party, an initial payment of $150 is required in addition to $15 per guest. (Let y be the total cost and x be the number of guests.)

   (J) $y = 150x + 15$
   (K) $y = 15x + 150$
   (L) $y = -150x + 15$
   (M) $y = -15x + 150$

x	y
-2	-7
-1	-4
0	-1
1	2

3. The table shows four pairs of x and y values. What is true for all values shown in the table above?

   (A) $y = -x - 5$
   (B) $y = x - 1$
   (C) $y = 2x + 1$
   (D) $y = 3x - 1$

4. Which of the following set of ordered pairs represents a function?

   (J) $\{(1, -2), (1, -1), (1, 0), (1, 1)\}$
   (K) $\{(-2, 3), (-1, 3), (0, 3), (-2, 5)\}$
   (L) $\{(1, 5), (2, 8), (3, 1), (4, -5)\}$
   (M) $\{(1, 2), (5, 3), (2, 3), (1, 5)\}$

5. Joshua is 5 years old and Jason is 3 times older. After 15 years, what will be the ratio of Jason's age to the Joshua's age?

(A) 3 : 2
(B) 2 : 3
(C) 2 : 1
(D) 4 : 3

6. When playing a certain game, Mr. Rhee always scores 12 less than twice of Joshua's score. Which of the following table represents this relationship?

(J)
Joshua	50	75	85
Rhee	118	168	188

(K)
Joshua	50	75	85
Rhee	88	138	158

(L)
Joshua	50	75	85
Rhee	88	168	158

(M)
Joshua	50	75	85
Rhee	118	138	188

7. A square has an area of 16. If the side length of the square increases by 25%, how much greater is the area of the new square compared to the area of the old?

(A) 9
(B) 8.5
(C) 8
(D) 7.5

8. If $y = \frac{3}{2}x + 1$ and $y = 7$, what is the value of x ?

(J) 3.5
(K) 4
(L) 4.5
(M) 5

$$\{(0,7), (1,6), (2,5), (3,4)\}$$

9. Which equation represents the relationship of the set of ordered pairs shown above?

(A) $y = -2x + 3$
(B) $y = -x + 7$
(C) $y = x + 3$
(D) $y = 2x + 7$

10. Which of the following is a pair of points for a line with a slope of $\frac{1}{2}$?

(J) $(2,3)$ and $(4,4)$
(K) $(2,3)$ and $(4,5)$
(L) $(2,3)$ and $(6,8)$
(M) $(2,3)$ and $(3,5)$

11. Which of the following table best represents the following verbal relationship? The number of jackets sold, y, is 4 less than three times the number of watches sold, x.

(A)
x	2	3	4	5
y	2	5	8	11

(B)
x	1	2	3	4
y	−1	3	6	9

(C)
x	0	1	2	3
y	4	7	10	13

(D)
x	1	2	3	4
y	2	5	8	11

12. If the x-intercept of $y = 3x + 6$ is -2, what are the xy-coordinates of the x-intercept?

(J) $(2, 0)$
(K) $(0, 2)$
(L) $(-2, 0)$
(M) $(0, -2)$

13. Which of the following line has a slope of zero?

(A) $y = 4$
(B) $x = 4$
(C) $y = 2x - 3$
(D) $y = -2x + 3$

14. Which of the following equation best represents the following verbal relationship? The average of x and y is 3.

(J) $2xy = 3$
(K) $\dfrac{xy}{2} = 3$
(L) $\dfrac{x+y}{2} = 3$
(M) $x + y = 3$

x	1	2	3
y	10	12	14

15. The table above contains ordered pairs that are solutions to which of the following equation?

(A) $y = 2x + 8$
(B) $y = x + 9$
(C) $y = -x + 11$
(D) $y = -2x + 12$

STOP

IAAT PRACTICE TEST 3

SECTION 4
Time — 10 minutes
15 Questions

Directions: Read the information given and choose the best answer for each question. Base your answer only on the information given. The time limit for each section is 10 minutes.

1. Joshua can read 25 pages in one hour. Which proportion can be used to determine r, the number of pages Joshua reads in 75 minutes?

    (A) $\dfrac{25}{1} = \dfrac{75}{r}$

    (B) $\dfrac{1}{25} = \dfrac{r}{75}$

    (C) $\dfrac{25}{60} = \dfrac{75}{r}$

    (D) $\dfrac{25}{60} = \dfrac{r}{75}$

2. Simplify: $4(-3x - 5)$

    (J) $-12x + 20$

    (K) $-12x - 5$

    (L) $12x + 5$

    (M) $-12x - 20$

3. If $3x + 2y = z$ and $z = 2y - 18$, what is the value of x?

    (A) -18

    (B) -6

    (C) 6

    (D) 18

4. If $a\Psi b = b^2 - 2a$, what is the value of $8\Psi 4$?

    (J) 0

    (K) 16

    (L) 32

    (M) 54

5. Solve for x: $-2(x + 4) = -12$

    (A) 4

    (B) 3

    (C) 2

    (D) 1

6. If $x = -4$ and $y = 6$, what is the value of $-x + xy$?

    (J) -20

    (K) -28

    (L) 20

    (M) 28

7. If $6x + 6y = 24$, what is the value of $x + y$?

   (A) 12
   (B) 8
   (C) 6
   (D) 4

Three less than four times the sum of a number, x, and 5

8. How is the following verbal phrase above expressed algebraically?

   (J) $3 - 4(\frac{x}{5})$
   (K) $4(x + 5) - 3$
   (L) $4(x - 5) - 3$
   (M) $3 - 4(x + 5)$

---

9. Simplify: $-4x + 2 + 6x - 8$

   (A) $2x + 6$
   (B) $2x - 6$
   (C) $-2x - 6$
   (D) $-2x + 6$

---

10. Solve for x: $\dfrac{9}{5} = \dfrac{45}{x}$

    (J) 25
    (K) 20
    (L) 12
    (M) 9

---

11. The volume of a cube is S^3, where S is the length of the cube. If the volume of a cube is 64, what is the length of the cube?

    (A) 4
    (B) 3
    (C) 2
    (D) 1

---

12. Which phrase is represented by $\frac{4}{3x} - 8$?

    (J) The product of four and three times a number, x, less eight.
    (K) Eight less than the quotient of four and a number, x.
    (L) The quotient of four and three times a number, x, less eight.
    (M) Eight less than three times a number, x, divided by four.

$$x - 3 < -4$$

13. What is the solution to the inequality above?

    (A) $x < 1$
    (B) $x > 1$
    (C) $x < -1$
    (D) $x > -1$

14. If $x = 4$ and $2x + 4y = -4$, what is the value of y?

   (J) -4
   (K) -3
   (L) 3
   (M) 4

15. Joshua was x years old 5 years ago from now. How old will he be 6 years from now?

   (A) $5x + 6$
   (B) $x + 11$
   (C) $x + 1$
   (D) $x - 1$

STOP

Answers and Solutions
IAAT Practice Test 3 Section 1

Answers

1. D	2. K	3. A	4. K	5. B
6. L	7. C	8. L	9. C	10. L
11. A	12. K	13. A	14. M	15. D

Solutions

1. (D)

   Use long division. $3630 \div 30 = 121$.

2. (K)

   Snowboard Waxing Company charges $7 every 100 square inches to wax a snowboard. The four snowboards, each with an area of 300 square inches, has a total area of $300 \times 4 = 1200$ square inches. The amount needed to be waxed is $1200 \div 100 = 12$ times the charged rate of work. Therefore, the total cost of waxing these four snowboards is $\$7 \times 12 = \$84$.

3. (A)

   To divide fractions, convert the number you are dividing by to the multiplicative inverse, or reciprocal, and then multiply. The multiplicative inverse, or reciprocal, of $\frac{7}{9}$ is $\frac{9}{7}$. Therefore,
   $$49 \div \frac{7}{9} = 49 \times \frac{9}{7} = 63$$

4. (K)

   A tennis player wins on the average 7 out of 12 matches. This means that he loses 5 out of 12 matches. Set up a proportion in order to determine the number of loses if he plays 72 matches.

   $\dfrac{\text{Loses}}{\text{Total matches}}$: $\dfrac{5}{12} = \dfrac{x}{72}$ (Cross multiply)

   $12x = 360$ (Divide each side by 12)

   $x = 30$

   Therefore, if the tennis player played 72 matches, he lost 30 matches.

SOLOMON ACADEMY
Distribution or replication of any part of this page is prohibited.

TEST 3 SECTION 1

5. (B)

   When dividing decimals, it is necessary to convert the number you are dividing by into a whole number. Convert 151.5 and 1.5 into 1515 and 15 respectively by shifting the decimal point of both numbers one place to the right. Afterwards, proceed using long division. Therefore,

   $$\frac{151.5}{1.5} = \frac{1515}{15} = 101$$

6. (L)

   Scientific notation must be written in the form: $c \times 10^n$, where $1 \leq c < 10$ and n must be integer. In general, positive n gives larger value than 10 and negative n gives smaller value than 1. Since $40,000,000 = 4 \times 10,000,000$, $40,000,000$ written in scientific notation is 4×10^7.

7. (C)

   Between 7:40 pm and 4:40 am of the next morning, 9 hours have passed. Lastly, in between 4:40 am and 5:10 am, 30 minutes have passed. Therefore, 9 hours and 30 minutes are in between 7:40 pm and 5:10 am of the next morning.

8. (L)

   When adding decimal numbers, it is necessary to line up the decimal points prior to adding. Make sure to add properly and carry over any necessary digits.

   $$\begin{array}{r} 2.1 \\ 0.3001 \\ +0.089 \\ \hline 2.4891 \end{array}$$

9. (C)

   It is possible to solve for the Greatest Common Factor (GCF) by listing out the factors for each number or by prime factorization.

   Factors of 56: 1, 2, 4, 7, 8, 14, **28**, and 56
   Factors of 84: 1, 2, 3, 4, 6, 7, 12, 14, 21, **28**, 42, and 84
   Since 28 is the largest factor of both 56 and 84, 28 is the GCF.

   OR

   The prime factorization of 56: **2 × 2 × 2 × 7**
   The prime factorization of 84: **2 × 2 × 3 × 7**
   Since there are at least two 2s and one 7 in both the prime factorization of 56 and 84, the GCF is $2 \times 2 \times 7 = 28$.

10. (L)

    A rectangle has a width of 7 and the length is 50% more than the width. This means that the length is $0.5 \times 7 = 3.5$ more than the width. Thus, the length is $7 + 3.5 = 10.5$. Therefore, the perimeter of the rectangle is $7 + 7 + 10.5 + 10.5 = 35$.

11. (A)

Since you have 5 types of bread, 3 types of meat, and 2 types of cheese, you can make $5 \times 3 \times 2 = 30$ different possible sandwiches if you use 1 type of bread, 1 type of meat, and 1 type of cheese.

12. (K)

To multiply decimal numbers, simply multiply ignoring the decimal points first. Afterwards, place the decimal point into the solution equivalent to the amount of decimal places the two numbers have together. Thus, change 0.56×0.23 into $56 \times 23 = 1288$. Since 0.56 has two numbers have the decimal point and 0.23 has two numbers after the decimal point, the solution will contain four decimal places. Therefore, $0.56 \times 0.23 = 0.1288$.

13. (A)

Let x be the price of a burger. The price of a soda is 60% of the price of a burger, which can be expressed as $0.6x$. Thus, the total price of a burger and a soda can be expressed as $x + 0.6x$ or $1.6x$. Since Mr. Rhee paid \$8 for a soda and burger, set $1.6x$ equal to \$8 and solve for x.

$$1.6x = 8 \qquad \text{(Divide each side by 1.6)}$$
$$x = 5.$$

Thus, the price of a burger is \$5. Since the price of a soda is 60% of the price of a burger, the price of a soda is $0.6x = 0.6 \times \$5 = \$3$.

14. (M)

In a book that contains 45 pages, there will be eleven 4's that appear on the page numbers of the book: 4, 14, 24, 34, 40, 41, 42, 43, 44, 45. Do not forget to count 44 twice because the digit 4 appears in both the tens and ones place. Therefore, (M) is the correct answer.

15. (D)

When evaluating mathematical expressions, use the order of operations: PEMDAS. This means solve expressions in the parenthesis first, then exponents, then multiplication and division from left to right, and finally addition and subtraction from left to right. Therefore,

$$(-2)^2 - 4(-3) = 4 - 4(-3)$$
$$= 4 + 12$$
$$= 16$$

SOLOMON ACADEMY — TEST 3 SECTION 2

Answers and Solutions
IAAT Practice Test 3 Section 2

Answers

1. A	2. L	3. A	4. L	5. B
6. J	7. C	8. M	9. A	10. K
11. B	12. L	13. D	14. M	15. A

Solutions

1. (A)

   The sum of grapes consumed by three people is 15. Since the mean is obtained by dividing the sum by 3, the mean number of grapes consumed by three people is $\frac{15}{3} = 5$.

2. (L)

   By observing the stacked bar graph, determine the number of fruits Sue consumed for each fruit. Sue had 1 apple, 6 oranges, 5 cherries, 2 grapes, and 3 peaches. Therefore, the fruit that Sue consumed the most is oranges.

3. (A)

   By observing the stacked bar graph, determine the number of fruits Jason consumed for each fruit. Jason had 3 apples, 3 oranges, 2 cherries, 4 grapes, and 4 peaches. Therefore, the fruit that Jason consumed the least is cherries.

4. (L)

   Joshua had 5 apples, 4 oranges, 1 cherry, 9 grapes, and 3 peaches. Therefore, Joshua consumed a total of $5 + 4 + 1 + 9 + 3 = 22$ fruits.

5. (B)

   Observe the graph in order to determine how many people were surveyed in all. There are $30 + 15 = 45$ people who favored science, $20 + 45 = 65$ people who favored math, $50 + 35 = 85$ people who favored English, $35 + 60 = 95$ people who favored history, and $65 + 45 = 110$ people who favored the arts. Therefore, the total number of people surveyed is $45 + 65 + 85 + 95 + 110 = 400$.

6. (J)

   Observe just the English portion of the bar graph. There were 50 7th graders and 35 8th graders who preferred English as their favorite subject. Therefore, there were $50 - 35 = 15$ more 7th graders than 8th graders that preferred English.

SOLOMON ACADEMY
Distribution or replication of any part of this page is prohibited.

TEST 3 SECTION 2

7. (C)

   Among the 7th and 8th graders, 110 students chose arts as their favorite subject. Therefore, the percent of students preferred is $\frac{110}{400} = 0.275 = \$27.5\%$.

8. (M)

   There were $35 + 60 = 95$ students who preferred History class and $50 + 35 = 85$ students who preferred English class. Therefore, there were $95 - 85 = 10$ more students who preferred history over English as their favorite subject.

9. (A)

   17 of the 50 people surveyed preferred blue as their favorite color. Therefore, the percentage of people who chose blue as their favorite color is $\frac{17}{50} = \frac{34}{100} = 0.34 = 34\%$.

10. (K)

    Front   **Back**
    1, 2, 3, 4, 5, 6, Joshua, 8, 9, 10, 11, 12, Jason, 14, 15

    Joshua is the 7th from the front and Jason is 3rd from the back as shown above. Since Joshua and Jason are not counted, the number of students between Joshua and Jason is 5.

11. (B)

    The total number of books sold on Monday, Tuesday, Thursday, and Friday is $22 + 52 + 17 + 33 = 124$. Since 189 books were sold during the week, the number of books sold on Wednesday is $189 - 124 = 65$.

12. (L)

    There are four soccer teams A, B, C, and D and each team plays against every other team only once. Therefore, there are 6 possible games between four soccer teams: AB, AC, AD, BC, BD, and CD.

13. (D)

    Assume that all of the pears and apples respectively weigh the same. 3 pears and 2 apples collectively weigh 29 ounces and 2 pears and 2 apples collectively weigh 24 ounces. Thus, in order to find the weight of a pear, subtract the 2nd equation from the 1st equation as shown below.

    $$\begin{array}{rl} 3 \text{ pears} + 2 \text{ apples} = & 29 \\ 2 \text{ pears} + 2 \text{ apples} = & 24 \\ \hline 1 \text{ pear} \phantom{+ 2 \text{ apples}} = & 5 \end{array}$$ Subtract two equations

    Therefore, the weight of a pear is 5 ounces.

14. (M)

Let x be the total number of students responded. Thus, 18% of all students responded can be expressed as $0.18x$. Since 54 students selected reading as their favorite extra-curricular activity, set $0.18x$ equal to 54 and solve for x.

$$0.18x = 54 \qquad \text{(Divide each side by 0.18)}$$
$$x = \frac{54}{0.18}$$
$$x = 300$$

Therefore, the total number of students responded to Jason's survey is 300.

15. (A)

Since Mr. Rhee runs only 5 days a week, he runs 10 days in two weeks. During those 10 days, Mr. Rhee runs a total of 90 miles. Therefore, Mr. Rhee runs $\frac{90}{10} = 9$ miles per day.

SOLOMON ACADEMY Distribution or replication of any part of this page is prohibited. **TEST 3 SECTION 3**

Answers and Solutions
IAAT Practice Test 3 Section 3

Answers

1. C	2. K	3. D	4. L	5. A
6. K	7. A	8. K	9. B	10. J
11. A	12. L	13. A	14. L	15. A

Solutions

1. (C)

   Since the points on the scatter plot fall from left to right, it shows a negative relationship.

2. (K)

   In order to describe the situation, let's use a linear equation model, $y = mx + b$, where m is the slope and b is the y-intercept. An initial payment of $150 is required to make a reservation for the party. Additionally, the cost per guest is $15. The initial payment of $150 is the y-intercept of the linear model because it is the fixed cost. Additionally, the total cost increases by 15 as the number of guest increases by 1. Thus, 15 is the slope of the linear model. Therefore, the equation that represents the situation is $y = 15x + 150$.

3. (D)

x	$y = 3x - 1$
-2	$y = 3(-2) - 1 = -7$
-1	$y = 3(-1) - 1 = -4$
0	$y = 3(0) - 1 = -1$
1	$y = 3(1) - 1 = 2$

   Plug in the x values into the equations listed as answer choices to determine which would yield the desired y values. The equation $y = 3x - 1$ is the only equation that satisfies the four ordered pairs $(-2, -7)$, $(-1, -4)$, $(0, -1)$, and $(1, 2)$.

4. (L)

   A function relates an input to an output such that x cannot have more than one value for an output y. In answer choices (J), (K), and (M), the x value repeats at least once. Thus, they do not represent a function and are incorrect. Answer choice (L) is a function by definition since all of the x values have exactly one y value. Therefore, (L) is the correct answer.

5. (A)

Joshua is 5 years old and Jason is 3 times older. In other words, Jason is $5 \times 3 = 15$ years old. In 15 years, Jason is 30 years old and Joshua is 20. Thus, the ratio of Jason's age to Joshua's age is 30:20. This ratio can be simplified by dividing by the greatest common factor which is 10. Therefore, the simplified ratio of Jason's age to Joshua's age is 3:2.

6. (K)

x	$y = 2x - 12$
50	$y = 2(50) - 12 = 88$
75	$y = 2(75) - 12 = 138$
85	$y = 2(85) - 12 = 158$

Mr. Rhee's always scores 12 less than twice of Joshua's score. Let y be Mr. Rhee's score and x be Joshua's score. Then, an equation relating the two scores can be expressed as $y = 2x - 12$ as shown above. Therefore, (K) is the correct answer.

7. (A)

A square has an area of 16. Since the sides of a square are all congruent, this means that the side of the square has a length of 4. The side length of the square increases by 25%. Since 25% of 4 is 1, the new side length of the square is $4 + 1 = 5$. Thus, the area of the new square is $5 \times 5 = 25$. Therefore, how much greater the area of the new square is compared to the area of the old is $25 - 16 = 9$.

8. (K)

Substitute 7 for y in the first equation. In order to solve for x in an equation, use the reverse order of operations, SADMEP, and inverse operations.

$$y = \frac{3}{2}x + 1 \quad \text{(Substitute 7 for } y\text{)}$$

$$7 = \frac{3}{2}x + 1 \quad \text{(Subtract each side by 1)}$$

$$6 = \frac{3}{2}x \quad \text{(Multiply each side by } \frac{2}{3}\text{)}$$

$$x = 4$$

Therefore, the value of x is 4.

9. (B)

x	$y = -x + 7$
0	$y = -(0) + 7 = 7$
1	$y = -(1) + 7 = 6$
2	$y = -(2) + 7 = 5$
3	$y = -(3) + 7 = 4$

Plug in the x values into the equations listed as answer choices to determine which would yield the desired y values. The equation $y = -x + 7$ is the only equation that satisfies the four ordered pairs $(0, 7)$, $(1, 6)$, $(2, 5)$, and $(3, 4)$.

10. (J)

A slope of $\frac{1}{2}$ means that the line will go up one step as it goes right two steps. The only pair of ordered pairs that show this pattern is answer choice (J). It is also possible to determine the slope by using the slope definition, $m = \frac{y_2 - y_1}{x_2 - x_1} = \frac{4-3}{4-2} = \frac{1}{2}$. Therefore, (J) is the correct answer.

11. (A)

x	$y = 3x - 4$
2	$y = 3(2) - 4 = 2$
3	$y = 3(3) - 4 = 5$
4	$y = 3(4) - 4 = 8$
5	$y = 3(5) - 4 = 11$

The number of jackets sold, y, is 4 less than three times the number of watches sold, x. This verbal relationship can be expressed as the equation $y = 3x - 4$. All ordered pairs in each table must satisfy this equation to be a solution. Make sure to test out all ordered pairs in a given table and do not assume that the other ordered pairs satisfy the equation. For example, the first ordered pair in answer choice (B) may lead you to believe that the table satisfies the equation because $(1, -1)$ is a solution. However, the second ordered pair $(2, 3)$ in answer choice (B) does not satisfy the equation. Thus, the entire table is invalid. The only table that satisfies $y = 3x - 4$ is answer choice (A) as represented in the table above. Therefore, (A) is the correct answer.

12. (L)

Since the x-intercept of the line is -2, the xy-coordinates of the x-intercept are $(-2, 0)$. Therefore, (L) is the correct answer.

13. (A)

Vertical lines (for example, $x = 4$) have undefined slope. Whereas, horizontal lines (for example, $y = 4$) have zero slope. Therefore, (A) is the correct answer.

14. (L)

The average of two numbers is sum of the two numbers divided by 2. Thus, the average of x and y is 3 can be expressed as $\dfrac{x+y}{2} = 3$. Therefore, (L) is the correct answer.

15. (A)

x	$y = 2x + 8$
1	$y = 2(1) + 8 = 10$
2	$y = 2(2) + 8 = 12$
3	$y = 2(3) + 8 = 14$

Plug in the x values into the equations listed as answer choices to determine which would yield the desired y values. The equation $y = 2x + 8$ is the only equation that satisfies all of the ordered pairs $(1, 10)$, $(2, 12)$ and $(3, 14)$.

SOLOMON ACADEMY Distribution or replication of any part of this page is prohibited. **TEST 3 SECTION 4**

Answers and Solutions
IAAT Practice Test 3 Section 4

Answers

1. D	2. M	3. B	4. J	5. C
6. J	7. D	8. K	9. B	10. J
11. A	12. L	13. C	14. K	15. B

Solutions

1. (D)

    One hour is equal to 60 minutes. Set up a proportion in terms of pages and minutes.

    $$25 \text{ pages} : 60 \text{ minutes} = r \text{ pages} : 75 \text{ minutes}$$
    $$\frac{25}{60} = \frac{r}{75}$$

    Therefore, (D) is the correct answer.

2. (M)

    Simplify the given expression by using the distributive property: $a(b+c) = ab + ac$. Thus,

    $$4(-3x - 5) = 4(-3x + -5) = 4(-3x) + 4(-5) = -12x - 20$$

    Therefore, (M) is the correct answer.

3. (B)

    Substitute $2y - 18$ for z in the first equation $3x + 2y = z$.

    $$\begin{aligned} 3x + 2y &= z & &\text{(Substitute } 2y - 18 \text{ for } z\text{)} \\ 3x + 2y &= 2y - 18 & &\text{(Subtract } 2y \text{ from each side)} \\ 3x &= -18 & &\text{(Divide each side by 3)} \\ x &= -6 \end{aligned}$$

    Therefore, the value of x is -6.

4. (J)

    $a \Psi b$ is defined as $b^2 - 2a$. In order to evaluate $8\Psi 4$, substitute 4 for b and substitute 8 for a.

    $$8\Psi 4 = 4^2 - 2(8) = 16 - 16 = 0$$

    Therefore, the value of $8\Psi 4$ is 0.

5. (C)

In order to solve the equation, divide each side of the equation by -2 and then solve for x.

$$-2(x+4) = -12 \quad \text{(Divide each side by } -2\text{)}$$
$$x + 4 = 6 \quad \text{(Subtract 4 from each side)}$$
$$x = 2$$

Therefore, the solution to $-2(x+4) = -12$ is $x = 2$.

6. (J)

Substitute -4 and 6 for x and y respectively and evaluate the expression. When two variables are next to each other without an operation, use multiplication. Plug-in negative values with a parenthesis to avoid confusion.

$$-x + xy = -(-4) + (-4)(6)$$
$$= 4 - 24$$
$$= -20$$

Therefore, the value of $-x + xy$ is -20.

7. (D)

In order to find the value of $x + y$, divide $6x + 6y = 24$ by 6.

$$6x + 6y = 24 \quad \text{(Divide each side by 6)}$$
$$\frac{6x + 6y}{6} = \frac{24}{6} \quad \text{(Split the expression on the left side)}$$
$$\frac{6x}{6} + \frac{6y}{6} = 4 \quad \text{(Simplify)}$$
$$x + y = 4$$

Therefore, the value of $x + y$ is 4.

8. (K)

Convert the verbal phrase into a mathematic equation.

Verbal Phrase	Expression
The sum of a number, x, and 5	$x + 5$
Four times the sum of a number, x and 5	$4(x + 5)$
Three less than four times the sum of a number, x and 5	$4(x + 5) - 3$

Therefore, (K) is the correct answer.

SOLOMON ACADEMY

Distribution or replication of any part of this page is prohibited.

TEST 3 SECTION 4

9. **(B)**

   In order to simplify the expression $-4x+2+6x-8$, combine like terms. $-4x$ and $6x$ are like terms because each term consists of x and has the same exponent. Thus, $-4x+6x$ can be combined as $-4x+6x=2x$. Since 2 and -8 are both constants they can be simplified as $2-8=-6$.

   $$-4x+2+6x-8 = (-4x+6x)+(2-8) = 2x-6$$

   Therefore, (B) is the correct answer.

10. **(J)**

    $$\frac{9}{5} = \frac{45}{x} \qquad \text{(Cross multiply)}$$
    $$9x = 45 \times 5 \qquad \text{(Divide each side by 9)}$$
    $$x = \frac{45 \times 5}{9} \qquad \text{(Simplify)}$$
    $$x = 25$$

    Therefore, the solution to $\frac{9}{5} = \frac{45}{x}$ is 25.

11. **(A)**

    The volume of the cube is 64. Since $64 = 4 \times 4 \times 4 = 4^3$, the length of the cube is 4. Therefore, (A) is the correct answer.

12. **(L)**

Verbal Phrase	Expression
Three times a number, x	$3x$
The quotient of four and three times a number, x	$\frac{4}{3x}$
The quotient of four and three times a number, x, less eight	$\frac{4}{3x} - 8$

    Therefore, (L) is the correct answer.

13. **(C)**

    In order to solve for x in an inequality, use the reverse order of operations, SADMEP, and inverse operations.

    $$x - 3 < -4 \qquad \text{(Add 3 to each side)}$$
    $$x < -1$$

    Therefore, the solution to $x - 3 < -4$ is $x < -1$.

SOLOMON ACADEMY

TEST 3 SECTION 4

14. (K)

Substitute 4 for x in the equation $2x + 4y = -4$ and solve for y. In order to solve for y, use the reverse order of operations, SADMEP, and inverse operations.

$$\begin{aligned} 2x + 4y &= -4 & &\text{(Substitute 4 for } x\text{)} \\ 2(4) + 4y &= -4 & &\text{(Subtract each side by 8)} \\ 4y &= -12 & &\text{(Divide each side by 4)} \\ y &= -3 \end{aligned}$$

Therefore, the value of y is -3.

15. (B)

Joshua was x years old 5 years ago from now, which implies that he is $x + 5$ years old now. Therefore, in 6 years from now, Joshua will be $x + 5 + 6$ or $x + 11$ years old.

IAAT PRACTICE TEST 4
SECTION 1
Time — 10 minutes
15 Questions

Directions: Read the information given and choose the best answer for each question. Base your answer only on the information given. The time limit for each section is 10 minutes.

1. The low temperature at night was 85°F. Each night for the next three nights, the low temperature dropped 6°F than the previous night. What was the low temperature of the last night?

    (A) 82°F

    (B) 79°F

    (C) 73°F

    (D) 67°F

2. Joshua has $\frac{3}{5}$ cup of milk. If the recipe called for $2\frac{3}{5}$ cups of milk, how many more cups of milk does he need?

    (J) 1

    (K) $1\frac{2}{5}$

    (L) 2

    (M) $2\frac{2}{5}$

2, 4, 8, 16, $\cdots$

3. What is the 6th number in the sequence above?

    (A) 24

    (B) 36

    (C) 48

    (D) 64

4. There are 1 red and 2 blue cards. What is the probability that a blue card is selected at random?

    (J) $\frac{3}{4}$

    (K) $\frac{2}{3}$

    (L) $\frac{1}{2}$

    (M) $\frac{1}{3}$

5. The price of a dozen eggs is $3.00. If the price is discounted 25%, what is the new price of the dozen eggs?

   (A) $2.75
   (B) $2.50
   (C) $2.25
   (D) $2.00

6. Which of the following is true?

   (J) $\frac{4}{100} = 40\%$
   (K) $0.08 = \frac{4}{50}$
   (L) $0.4 = \frac{0.4}{100}$
   (M) $\frac{4}{50} = 80\%$

7. While baking, Sue needs 2 tablespoons of sugar for every 7 quarts of milk. How many tablespoons of sugar would she need for 4 quarts of milk?

   (A) $\frac{8}{7}$
   (B) $\frac{7}{8}$
   (C) $\frac{7}{4}$
   (D) $\frac{4}{7}$

8. Use the following information to determine who received the lowest percent of getting right answers. For every 10 questions, Sue answered 9 correctly. Joshua answered 44 out of 50 questions correctly. Out of 100 questions, Mr. Rhee answered 8 questions incorrectly. Jason received 91% of getting right answers.

   (J) Jason
   (K) Sue
   (L) Joshua
   (M) Mr. Rhee

9. The number of boys in the class is twice as many as the number of girls. If there are 72 students, how many are boys?

   (A) 60
   (B) 48
   (C) 36
   (D) 24

10. Jason buys a pack of 8-batteries for $2.45. Which of the following is closest to the unit price of a battery?

    (J) $0.31
    (K) $0.36
    (L) $0.42
    (M) $0.46

11. If Sue is driving at a rate of 13 miles per 15 minutes, what is her rate in terms of miles per hour?

   (A) 62
   (B) 60
   (C) 55
   (D) 52

12. For a party, Mr. Rhee ordered a 5-foot sub. If the sub was cut in 15 equal pieces, how many inches long will each piece be? (1 foot equals 12 inches.)

   (J) 7
   (K) 6
   (L) 5
   (M) 4

13. $\dfrac{5^3 \times 5^2}{5^7} =$

   (A) 5^3
   (B) 5^2
   (C) 5^{-2}
   (D) 5^{-3}

14. Which of the following inequality is true?

   (J) $3.54 \times 10^1 < 3.54 \times 10^{-1}$
   (K) $1.65 \times 10^{-1} > 1.65 \times 10^{-2}$
   (L) $1.35 \times 10^3 < 1.35 \times 10^2$
   (M) $9.81 \times 10^{-5} > 9.81 \times 10^{-4}$

15. Evaluate: $\dfrac{1}{4} + \dfrac{11}{12} - \dfrac{2}{3}$

   (A) $-\dfrac{1}{2}$
   (B) $\dfrac{1}{2}$
   (C) $\dfrac{1}{3}$
   (D) $\dfrac{2}{3}$

STOP

IAAT PRACTICE TEST 4

SECTION 2
Time — 10 minutes
15 Questions

Directions: Read the information given and choose the best answer for each question. Base your answer only on the information given. The time limit for each section is 10 minutes.

Directions: Use the following graph to answer questions 1 – 4.

Survey of Favorite Colors

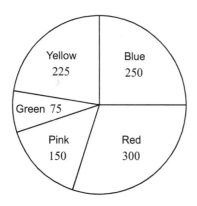

1. What percent of those surveyed favored the color red?

    (A) 30%
    (B) 35%
    (C) 40%
    (D) 45%

2. Which color was the least favorite?

    (J) Blue
    (K) Red
    (L) Pink
    (M) Green

3. Which two colors when added together has the same number as those who favored the color red?

    (A) Yellow and Pink
    (B) Blue and Pink
    (C) Yellow and Green
    (D) Green and Blue

4. According to the pie chart, what is the probability that a participant chosen at random would favor the color blue?

    (J) 0.25
    (K) 0.3
    (L) 0.35
    (M) 0.4

Directions: Use the following graph to answer questions 5 – 8.

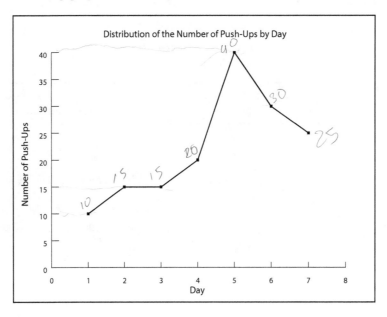

5. How many push-ups were completed in all?

   (A) 140
   (B) 145
   (C) 150
   (D) 155

6. Estimate the mean number of push-ups completed from Day 1 to Day 7, inclusive.

   (J) 18
   (K) 22
   (L) 26
   (M) 30

7. What is the median amount of push-ups completed?

   (A) 15
   (B) 20
   (C) 25
   (D) 30

8. What is the mode number of push-ups completed?

   (J) 25
   (K) 20
   (L) 15
   (M) There is no mode.

9. A dentist goes into his dental office at 7 am and sees patients until his hour long lunch break at 12 pm. After lunch, he continues to see patients until 3 pm. After taking a 15-minute break, he continues to see patients for an additional two hours starting at 3:15 pm. How many hours did the dentist see patients in his office?

(A) 9 Hours and 45 Minutes

(B) 9 Hours and 30 Minutes

(C) 9 Hours and 15 Minutes

(D) 9 Hours

10. Mr. Rhee, Sue, Jason, and Joshua are standing in a line from left to right: Jason is not the first person from left. Sue is not first person from the right. Jason is standing left of Joshua. Someone is standing right of Joshua. List the order from left to right.

(J) Sue, Jason, Joshua, Mr. Rhee

(K) Jason, Joshua, Mr. Rhee, Sue

(L) Jason, Joshua, Sue, Mr. Rhee

(M) Mr. Rhee, Jason, Sue, Joshua

Bicycling	♡ ♡ ♡
Running	♡ ♡ ♡ ♡ ♡
Rowing	♡ ♡

♡ = 700 Calories

11. Mr. Rhee is going to a gym to lose weight. The table above summarizes the types of cardiovascular exercises he had done in November. Assuming 3500 calories is equal to 1 pound of body fat, how many pounds of body fat did Mr. Rhee lose in November?

(A) 4

(B) 3

(C) 2

(D) 1

{4, 24, 3, 5, 9, 2, 18}

12. What is the range of the data set shown above?

(J) 24

(K) 22

(L) 14

(M) 2

SOLOMON ACADEMY — Distribution or replication of any part of this page is prohibited. — TEST 4 SECTION 2

{3, 7, 1, 11, 2, 15}

13. What is the median of the data set shown above?

 (A) 6
 (B) 3
 (C) 5
 (D) 7

A B C D A B ⋯

14. If the pattern above is repeated, what is the symbol of the 13th term?

 (J) A
 (K) B
 (L) C
 (M) D

15. A clock is malfunctioning. The minute hand of the clock only moves and indicates correct time every 12 minutes. For instance, the clock indicates 12 pm between 12 pm to 12:11 pm and indicates correct time at 12:12 pm. How many times does the clock indicate correct time between 12:10 pm and 1:10 pm?

 (A) 8 times
 (B) 7 times
 (C) 6 times
 (D) 5 times

STOP

IAAT PRACTICE TEST 4

SECTION 3
Time — 10 minutes
15 Questions

Directions: Read the information given and choose the best answer for each question. Base your answer only on the information given. The time limit for each section is 10 minutes.

1. The table below shows four pairs of x and y values. What is true for all values shown in the table above?

x	y
-1	4
0	1
1	-2
2	-5

   (A) No Relationship
   (B) $y = x + 5$
   (C) $y = -x + 3$
   (D) $y = -3x + 1$

2. You agree to finance a $1800 computer for a year by paying monthly. What is your monthly payment for the computer?

   (J) $100
   (K) $125
   (L) $150
   (M) $175

3. Jason has 13 marbles in the beginning and collects 3 marbles per day. Which of the following equation represents the number of marbles, y, that Jason has after x days?

   (A) $y = 13x - 3$
   (B) $y = 13x + 3$
   (C) $y = 3x - 13$
   (D) $y = 3x + 13$

4. Which of the following set of ordered pairs represents the equation, $y = 5 - x$?

   (J) $\{(1,4), (2,3), (3,2), (4,1)\}$
   (K) $\{(5,10), (4,9), (3,8), (0,5)\}$
   (L) $\{(-1,5), (-2,6), (-3,7), (-4,8)\}$
   (M) $\{(-2,7), (-1,6), (0,5), (1,4)\}$

5. Jason has 3 less than twice as many points as Joshua. How many points does Jason have if Joshua has 6 points?

   (A) 3
   (B) 6
   (C) 9
   (D) 12

x	1	2	3	4	5
y	3	7	11	15	

6. Observe the numbers in the two columns shown above. Which of the following belongs in the empty cell.

   (J) 20
   (K) 19
   (L) 18
   (M) 17

7. The area of a triangle is $\frac{1}{2}bh$, where b is the base and h is the height. If the lengths of the base and height are 13 and 2, respectively, what is the area of the triangle?

   (A) 12
   (B) 13
   (C) 14
   (D) 15

8. If ℓ and w represent the length and width of a rectangle, which of the following can be used to determine the perimeter of a rectangle?

   (J) $\ell \times w$
   (K) $\ell + w$
   (L) $\frac{\ell+w}{2}$
   (M) $2(\ell+w)$

The number of cups of coffee sold, y, is two more than the number of pastries sold, x.

9. Which table contains only values that satisfy the following verbal phrase above?

   (A)

   (B)

   (C)

   (D)
x	3	5	7	9
y	6	10	14	18

10. What is the x-intercept of the equation $2y + 3x = 2$?

    (J) $\frac{3}{2}$
    (K) $\frac{2}{3}$
    (L) 2
    (M) 3

11. Which of the following equation best represents the following verbal relationship? The number of carrots, c, is three more than eight times the number of beets, b.

    (A) $c = 3b + 8$
    (B) $c = 8b + 3$
    (C) $b = 3c + 8$
    (D) $b = 8c + 3$

12. If $y = 4x - 5$ and $x = 2$, what is the value of y?

    (J) 3
    (K) 1
    (L) -1
    (M) -3

13. Which of the following equation represents a line that is parallel to $y = -\frac{1}{2}x - 3$?

    (A) $y = 2x - 3$
    (B) $y = -2x + 3$
    (C) $y = \frac{1}{2}x + 3$
    (D) $y = -\frac{1}{2}x + 3$

14. Joshua loves to play basketball and makes 35% of the shots attempted. Which of the following is the equation used to determine how many basket Joshua makes, x, if he shoots 30 shots?

    (J) $x = \frac{30}{0.35}$
    (K) $x = \frac{30}{35}$
    (L) $x = 30 \times 0.35$
    (M) $x = 30 \times 35$

15. Mr. Rhee wants to buy a suit since the entire store is on sale 30%. The store charges a sales tax of 10% based on the price after all discounts are applied. What is the total cost Mr. Rhee will pay for a suit that originally cost x dollars?

    (A) $x(0.3)(0.1)$
    (B) $x(1 + 0.3)(1 - 0.1)$
    (C) $x(1 - 0.3)(1 + 0.1)$
    (D) $x(1 - 0.3)(1 + 10)$

STOP

IAAT PRACTICE TEST 4

SECTION 4
Time — 10 minutes
15 Questions

Directions: Read the information given and choose the best answer for each question. Base your answer only on the information given. The time limit for each section is 10 minutes.

1. If $2^x = 8$, what is the value of x^2?

   (A) 8
   (B) 9
   (C) 16
   (D) 25

2. $\dfrac{2x+4}{2} =$

   (J) $x+4$
   (K) $x+2$
   (L) $2x+2$
   (M) $2x+4$

3. x is twice as long as y. y is one-third the length of z. If $z = 24$, what is the value of x?

   (A) 8
   (B) 12
   (C) 16
   (D) 18

4. The area of a circle is defined as πr^2, where r represents the radius. What is the area of a circle that has a radius of 4?

   (J) 32π
   (K) 16π
   (L) 8π
   (M) 4π

5. Simplify: $4x + 2(2x - 3)$

   (A) $12x + 6$
   (B) $12x - 6$
   (C) $8x + 6$
   (D) $8x - 6$

6. Solve for x: $18 - 2x = 14$

   (J) -2
   (K) 2
   (L) 3
   (M) 4

7. If $x = 7$ and $y = -4$, what is the value of $2x - 3y$?

   (A) 2
   (B) 8
   (C) 12
   (D) 26

The sum of x and y is 10

8. How is the verbal phrase above expressed algebraically?

   (J) $xy = 10$
   (K) $\frac{y}{x} = 10$
   (L) $x + y = 10$
   (M) $y - x = 10$

---

9. x and y are positive integers. If $x + y = 10$ and $x - y = 2$, what are the values of x and y?

   (A) $x = 8$ and $y = 2$
   (B) $x = 2$ and $y = 8$
   (C) $x = 6$ and $y = 4$
   (D) $x = 4$ and $y = 6$

---

10. If $2x = 3y = 6$, what is the value of $x + y$?

    (J) 6
    (K) 7
    (L) 6
    (M) 5

---

11. What is the solution to $-2x + 1 > 5$?

    (A) $x < 2$
    (B) $x > 2$
    (C) $x < -2$
    (D) $x > -2$

---

12. The initial volume of a balloon is V. The volume of the balloon is doubled every three minutes. After 9 minutes, what is the new volume of the balloon in terms of V?

    (J) $8V$
    (K) $9V$
    (L) $10V$
    (M) $12V$

---

13. Joshua is making a scale model of his favorite helicopter. The actual helicopter is 40 feet long and 9 feet tall. Joshua wants to make his model to be 15 inches in length. Which could be used to find the height of his model if the same ratio is used?

    (A) $\frac{40}{x} = \frac{9}{15}$
    (B) $\frac{40}{15} = \frac{x}{9}$
    (C) $\frac{40}{9} = \frac{15}{x}$
    (D) $\frac{40}{9} = \frac{x}{15}$

14. If $x - 2 = -5$, what is the value of $\frac{6-3x}{3}$?

   (J) 5
   (K) 4
   (L) 3
   (M) 2

15. Solve for x: $\sqrt{x} - 1 = 3$

   (A) 16
   (B) 9
   (C) 4
   (D) 2

STOP

SOLOMON ACADEMY Distribution or replication of any part of this page is prohibited. TEST 4 SECTION 1

Answers and Solutions
IAAT Practice Test 4 Section 1

Answers

1. D	2. L	3. D	4. K	5. C
6. K	7. A	8. L	9. B	10. J
11. D	12. M	13. C	14. K	15. B

Solutions

1. (D)

    The low temperature at night was 85°F. Each night for the next three nights, the low temperature dropped 6°F than the previous night. Thus, the low temperatures for the next three nights were 79°F, 73°F, and 67°F. Therefore, (D) is the correct answer.

2. (L)

    Joshua has $\frac{3}{5}$ cup of milk and the recipe called $2\frac{3}{5}$ cups of milk. In order to determine the amount of milk that Joshua needs, proceed to subtract.

    $$2\frac{3}{5} - \frac{3}{5} = 2 + \frac{3}{5} - \frac{3}{5} = 2$$

    Therefore, Joshua needs 2 more cups of milk.

3. (D)

    The sequence, $2, 4, 8, 16, \cdots$, suggests a pattern such that the value of the next term is twice the value of the previous term.

    $$2, 4, 8, 16, 32, 64, \cdots$$

    Therefore, the 6th number of the sequence is 64.

4. (K)

    Out of the three cards, two cards are blue. Therefore, the probability of selecting a blue cards is $\frac{2}{3}$.

5. (C)

    The price of a dozen eggs is $3.00. If the price is discounted 25%, the new price is 75% of the original price. Therefore, the new price of the dozen eggs is $0.75 \times \$3.00 = \$2.25$.

6. (K)

    Since $0.08 = \frac{8}{100} = \frac{4}{50}$, (K) is the correct answer.

7. (A)

Set up a proportion in terms of tablespoons of sugar and quarts of milk to determine the amount of green paint needed when using 28 ounces of white paint.

$$2_{\text{sugar}} : 7_{\text{milk}} = x_{\text{sugar}} : 4_{\text{milk}}$$
$$\frac{2}{7} = \frac{x}{4} \qquad \text{(Cross multiply)}$$
$$7x = 8 \qquad \text{(Divide each side by 7)}$$
$$x = \frac{8}{7}$$

Therefore, Sue would need $\frac{8}{7}$ tablespoons of sugar for 4 quarts of milk.

8. (L)

For every 10 questions, Sue answered 9 correctly. This means that Sue received $\frac{9}{10} = \frac{90}{100} = 90\%$ of getting right answer. Since Joshua answered 44 out of 50 questions correctly, he received $\frac{44}{50} = \frac{88}{100} = 88\%$. Furthermore, out of 100 questions, Mr. Rhee answered 8 questions incorrectly, which means that Mr. Rhee answered $100 - 8 = 92$ questions correctly. Thus, he received $\frac{92}{100} = 92\%$. Lastly, Jason received 91%. Since the person who received the lowest percent of getting right answers is Joshua, (L) is the correct answer.

9. (B)

The number of boys in the class is twice as many as the number of girls can be represented by the ratio 2 : 1. That implies that the number of boys is $\frac{2}{3}$ of the total number of students, 72. Therefore, the number of boys in the class is $\frac{2}{3} \times 72 = 48$.

10. (J)

Jason buys a pack of 8-batteries for $2.45. In order to find the unit price of a battery, or the cost of a single battery, use division.

$$\frac{\$2.45}{8} = 0.30625 \approx 0.31$$

Therefore, the unit price of a battery is $0.31.

11. (D)

Sue is driving at a rate of 13 miles per 15 minutes. In order to determine her rate in terms of miles per hour, multiply the numerator and denominator by 4.

$$\frac{13 \text{ miles}}{15 \text{ minutes}} = \frac{13 \text{ miles} \times 4}{15 \text{ minutes} \times 4}$$
$$= \frac{52 \text{ miles}}{60 \text{ minutes}}$$
$$= \frac{52 \text{ miles}}{\text{hour}}$$

Therefore, Sue is driving at 52 miles per hour.

12. (M)

    Mr. Rhee ordered a 5-foot sub. The 5-foot sub is cut into 15 equal piece. In order to determine the length of each piece, convert 5 feet into inches: 5 feet = 60 inches. Therefore, the length of each piece after the cut is $\frac{60 \text{ inches}}{15} = 4$ inches.

13. (C)

    Recall the properties of exponents: $a^x \times a^y = a^{x+y}$ and $\frac{a^x}{a^y} = a^{x-y}$.

    $$\frac{5^3 \times 5^2}{5^7} = \frac{5^5}{5^7} = 5^{5-7} = 5^{-2}$$

    Therefore, (C) is the correct answer.

14. (K)

    When properly written in scientific notation, it is easy to compare numbers by observing the integer value of the exponent. When comparing two numbers, the exponent with the larger integer value will be greater in value. Since none of the comparing values in the answer choices have the same exponents, determine if the inequality is true by observing only the exponent. In answer choice (K), since -1 is greater than -2, the inequality is true. Therefore, (K) is the correct answer.

15. (B)

    When adding or subtracting fractions with uncommon denominators, it is necessary to have a common denominator. Since the least common multiple (LCM) of 4, 12, and 3 is 12, convert $\frac{1}{4} = \frac{3}{12}$, and $\frac{2}{3} = \frac{8}{12}$. Now, since all fractions have a common denominator, proceed to add and subtract.

    $$\frac{1}{4} + \frac{11}{12} - \frac{2}{3} = \frac{3}{12} + \frac{11}{12} - \frac{8}{12}$$
    $$= \frac{6}{12}$$
    $$= \frac{1}{2}$$

    Therefore, (B) is the correct answer.

SOLOMON ACADEMY — Distribution or replication of any part of this page is prohibited. — TEST 4 SECTION 2

Answers and Solutions
IAAT Practice Test 4 Section 2

Answers

1. A	2. M	3. C	4. J	5. D
6. K	7. B	8. L	9. D	10. J
11. C	12. K	13. C	14. J	15. D

Solutions

1. (A)

   The total number of those surveyed is $75 + 225 + 250 + 300 + 150 = 1000$. Of those 1000 surveyed, 300 people chose red as their favorite color. Therefore, the percent of those who favored the color red is $\frac{300}{1000} = \frac{30}{100} = 30\%$.

2. (M)

   Out of all the colors, only 75 people chose the color green. Therefore, the least favorite color is green.

3. (C)

   300 people chose the color red. Since 225 people chose the color yellow and 75 people chose the color green, the two colors add up to 300. Thus, those who favored either the color yellow or green is equal to those who favored the color red. Therefore, (C) is the correct answer.

4. (J)

   The total number of those surveyed who chose the colors blue is 250. Therefore, the probability that a participant chosen at random would favor the color blue is $\frac{250}{1000} = 0.25$.

5. (D)

   Add up the number of push-ups that were completed each day: $10 + 15 + 15 + 20 + 40 + 30 + 25 = 155$. Therefore, 155 push-ups were completed in all.

6. (K)

   The total number of push-ups that were completed from Day 1 to Day 7 is 155. Since you are estimating the mean number of push-ups, use 154 as the total number of push-ups. Thus, the mean number of push-ups is $\frac{154}{7} = 22$. Therefore, (K) is the correct answer.

SOLOMON ACADEMY

Distribution or replication of any part of this page is prohibited.

TEST 4 SECTION 2

7. (B)

$$\{10, 15, 15, 20, 25, 30, 40\}$$

As shown above, it is necessary to order the numbers from least to greatest when trying to find the median. Since the median is the middle number, the median amount of push-ups completed is 20.

8. (L)

The mode is a number that appears the most. Since 15 appear twice while all other numbers appear only once, the mode number of push-ups completed is 15.

9. (D)

The question is asking how many hours the dentist saw patients during this day. From 7 am to 12 pm, the dentist sees patients for a total of 5 hours. Since the dentist takes an hour long lunch break, he continues to see patients from 1 pm to 3 pm which is 2 hours. After a short break, the dentists continues to see patients for an additional 2 hours. Therefore, the total number of hours the dentist sees patients in his office is $5 + 2 + 2 = 9$ hours.

10. (J)

Since Jason is not the first person from left, answer choices (K) and (L) are immediately eliminated. Jason is standing left of Joshua. Therefore, answer choice (J) is the only one that satisfies this information and is the correct answer: Sue, Jason, Joshua, and Mr. Rhee.

11. (C)

According to the table, the total amount of calories that Mr. Rhee lost in November is $10 \times 700 = 7000$. Since 3500 calories is equal to 1 pound of body fat,

$$\text{Total pounds of body fat lost} = \frac{7000}{3500} = 2$$

Therefore, Mr. Rhee lost 2 pounds of body fat in November.

12. (K)

In order to determine the range of the set, subtract the largest value by the smallest value. Therefore, the range of the data set is $24 - 2 = 22$.

13. (C)

$$\{1, 2, 3, 7, 11, 15\}$$

The median is the middle number. It is necessary to rearrange the numbers from least to greatest when determining the median as shown above. Since the set has an even number of numbers, the median is the average of the two middle numbers. Therefore, the median of the set is $\frac{3+7}{2} = 5$.

14. (J)

The pattern consists of four symbols: A, B, C, D. This means that every fourth term, or multiple of four, will have the symbol D. Thus, the 4th, the 8th and the 12th are D. Therefore, the 13th is A.

15. (D)

Between 12:10 pm and 1:10 pm, the clock indicates the first correct time at 12:12pm. Afterwards, it indicates correctly at 12:24pm, 12:36pm, 12:48pm and 1:00pm. Therefore, the clock indicates time correctly 5 times between 12:10 pm and 1:10 pm.

SOLOMON ACADEMY — TEST 4 SECTION 3

Answers and Solutions
IAAT Practice Test 4 Section 3

Answers

1. D	2. L	3. D	4. J	5. C
6. K	7. B	8. M	9. A	10. K
11. B	12. J	13. D	14. L	15. C

Solutions

1. (D)

   Observe the different equations of answer choices to see which satisfies the set of ordered pairs. It is important to test all of the listed ordered pairs to avoid confusion. For example, answer choice (B) only satisfies the first listed ordered pair. Do not jump to conclusions but test out multiple ordered pairs. The equation that is true for all the listed ordered pairs in the table is answer choice (D), $y = -3x + 1$.

2. (L)

   There are 12 months in a year. Thus, the monthly payment of a $1800 computer is $\frac{\$1800}{12} = \$150$.

3. (D)

   Jason currently has 13 marbles in the beginning, which serves as the y-intercept. Jason continues to collect marbles at a rate of 3 marbles per day. In other words, Jason has collects $3x$ marbles in x days. Thus, the total number of marbles, y, that Jason has after x days can be represented by the equation $y = 3x + 13$.

4. (J)

(x,y)	$y = 5 - x$	Satisfy
$(1,4)$	$4 = 5 - 1$	✓
$(2,3)$	$3 = 5 - 2$	✓
$(3,2)$	$2 = 5 - 3$	✓
$(4,1)$	$1 = 5 - 4$	✓

   Plug in the different ordered pairs into the equation $y = 5 - x$. Answer choice (J) is a set of ordered pairs that completely satisfies the equation $y = 5 - x$.

5. (C)

Joshua has 6 points. Since Jason has 3 less than twice as many points as Joshua, Jason has $2 \times 6 - 3 = 9$ points. Therefore, (C) is the correct answer.

6. (K)

x	1	2	3	4	5
y	3	7	11	15	19

As the value of x increases by 1, the value of y increases by 4. Therefore, when the value of x is 5, the value of y is $15 + 4 = 19$.

7. (B)

The area of a triangle is $\frac{1}{2}bh$, where b is the base and h is the height. Since the lengths of the base and height are 13 and 2, respectively, the area of the triangle is $\frac{1}{2}(13)(2) = 13$.

8. (M)

ℓ and w represent the length and width of a rectangle. The perimeter of a rectangle is the distance around the rectangle. Thus,

$$\begin{aligned} \text{Perimeter of a rectangle} &= \ell + \ell + w + w \\ &= 2\ell + 2w \qquad \text{(Factor out a 2)} \\ &= 2(\ell + w) \end{aligned}$$

Therefore, (M) is the correct answer.

9. (A)

x	$y = x + 2$
1	$y = 1 + 2 = 3$
2	$y = 2 + 2 = 4$
3	$y = 3 + 2 = 5$
4	$y = 4 + 2 = 6$

The number of cups of coffee sold, y is two more than the number of pastries sold, x, can be represented by the equation $y = x + 2$. The table above shows that all the ordered pairs in answer choice (A) satisfy the equation.

10. (K)

In order to find the x-intercept, substitute 0 for y and solve for x.

$$\begin{aligned} 2y + 3x &= 2 \qquad \text{(Substitute 0 for } y\text{)} \\ 3x &= 2 \qquad \text{(Divide each side by 3)} \\ x &= \frac{2}{3} \end{aligned}$$

Therefore, the x-intercept of the equation $2y + 3x = 2$ is $\frac{2}{3}$.

11. (B)

Verbal Phrase	Expression
Eight times the number of beets, b	$8b$
Three more than eight times the number of beets, b	$8b + 3$

Since the number of carrots, c, is three more than eight times the number of beets, b, the equation that represents the verbal relationship is $c = 8b + 3$.

12. (J)

Substitute 2 for x in the equation $y = 4x - 5$ and solve for y.

$y = 4x - 5$ (Substitute 2 for x)
$y = 4(2) - 5$ (Simplify)
$y = 3$

Therefore, the value of y is 3.

13. (D)

The equation of the line is written in slope-intercept form: $y = mx + b$, where m and b represent the slope and y-intercept respectively. Two lines are considered parallel to each other if they have the same slope but different y-intercept. Since the slope of the line $y = -\frac{1}{2}x - 3$ is $-\frac{1}{2}$, the equation of the line that is parallel to $y = -\frac{1}{2}x - 3$ is $y = -\frac{1}{2}x + 3$ in answer choice (D). Therefore, (D) is the correct answer.

14. (L)

In order to determine the number of basket Joshua makes, multiply the total attempts by the percentage he makes. Convert 35% into decimal form by moving the decimal point two places to the left or by dividing by 100. Thus, $35\% = 0.35$. Therefore, the equation used to determine how many basket Joshua makes is $x = 30 \times 0.35$.

15. (C)

Mr. Rhee wants to buy a suit that is on sale 30%. Note that $30\% = 0.3$. The price that Mr. Rhee needs to pay for the suit is 70% or $(1 - 0.3)$ of the original price of the suit, x, which can be expressed as $x(1 - 0.3)$. The store charges a sales tax of 10%, which is 10% of $x(1 - 0.3)$ or $x(1 - 0.3)(0.1)$. Since the total cost consists of the price of the suit, $x(1 - 0.3)$, and a sales tax, $x(1 - 0.3)(0.1)$, the total cost that Mr. Rhee will pay for the suit is $x(1 - 0.3) + x(1 - 0.3)(0.1)$ or $x(1 - 0.3)(1 + 0.1)$. Therefore, (C) is the correct answer.

SOLOMON ACADEMY — TEST 4 SECTION 4

Answers and Solutions
IAAT Practice Test 4 Section 4

Answers

1. B	2. K	3. C	4. K	5. D
6. K	7. D	8. L	9. C	10. M
11. C	12. J	13. C	14. J	15. A

Solutions

1. (B)

    The exponent of a number tells you how many times you multiply the number by itself. Since $2 \times 2 \times 2 = 8$, $2^3 = 8$. Thus, $x = 3$. Therefore, the value of $x^2 = 3^2 = 9$.

2. (K)

    $$\frac{2x+4}{2} = \frac{2x}{2} + \frac{4}{2} = x + 2$$

    Therefore, (K) is the correct answer.

3. (C)

    x is twice as long as y can be expressed as $x = 2y$. Furthermore, y is one-third the length of z can be expressed as $y = \frac{1}{3}z$. Since $z = 24$, $y = \frac{1}{3}z = \frac{1}{3}(24) = 8$. Therefore, the value of x is $x = 2y = 2(8) = 16$.

4. (K)

    The area of a circle is defined as πr^2, where r represents the radius. Since the radius of a circle is 4, the area of the circle is $\pi(4)^2 = 16\pi$.

5. (D)

    In order to simplify the expression, use the distributive property: $a(b+c) = ab + ac$.

    $$4x + 2(2x - 3) = 4x + 4x - 6 = 8x - 6$$

    Therefore, (D) is the correct answer.

6. (K)

$$18 - 2x = 14 \qquad \text{(Subtract each side by 18)}$$
$$-2x = -4 \qquad \text{(Divide each side by -2)}$$
$$x = 2$$

Therefore, the solution to $18 - 2x = 14$ is $x = 2$.

7. (D)

Substitute 7 for x and -4 for y.

$$2x - 3y = 2(7) - 3(-4) = 14 + 12 = 26$$

Therefore, when $x = 7$ and $y = -4$, the value of $2x - 3y = 26$.

8. (L)

The sum of x and y is 10 can be expressed as $x + y = 10$. Therefore, (C) is the correct answer.

9. (C)

x and y are positive integers. The only values of x and y that satisfy the equations $x + y = 10$ and $x - y = 2$ are $x = 6$ and $y = 4$. Therefore, (C) is the correct answer.

10. (M)

Since $2x = 3y = 6$, $2x = 6$ and $3y = 6$. Thus, $x = 3$ and $y = 2$. Therefore, the value of $x + y = 3 + 2 = 5$.

11. (C)

When multiplying or dividing each side of the inequality by a negative number, the inequality symbol is reversed.

$$-2x + 1 > 5 \qquad \text{(Subtract 1 from each side)}$$
$$-2x > 4 \qquad \text{(Divide each side by } -2\text{)}$$
$$x < -2 \qquad \text{(Reverse the inequality symbol)}$$

Therefore, (C) is the correct answer.

12. (J)

The initial volume of a balloon is V. The volume of the balloon is doubled every three minutes.

Initial Volume	After 3 minutes	After 6 minutes	After 9 minutes
V	$2V$	$4V$	$8V$

Therefore, the volume of the balloon after 9 minutes is $8V$ as shown above.

13. (C)

Proportions state that two ratios, usually expressed as fractions, are the same. The proportion in answer choice (C) correctly represents the ratio of the length to the height.

$$\frac{\text{Length}}{\text{Height}} : \frac{40\,\text{feet}}{9\,\text{feet}} = \frac{15\,\text{inches}}{x\,\text{inches}}$$

Therefore, (C) is the correct answer.

14. (J)

Since $x - 2 = -5$, $x = -3$. Substitute -3 for x in the expression $\frac{6-3x}{3}$.

$$\frac{6 - 3x}{3} = \frac{6 - 3(-3)}{3} \quad \text{(Substitute } -3 \text{ for } x\text{)}$$
$$= \frac{15}{3}$$
$$= 5$$

Therefore, the value of $\frac{6-3x}{3}$ when $x = -3$ is 5.

15. (A)

In order to solve for x, use the reverse of PEMDAS: SADMEP.

$$\sqrt{x} - 1 = 3 \quad \text{(Add 1 to each side)}$$
$$\sqrt{x} = 4 \quad \text{(Square each side)}$$
$$x = 16$$

Therefore, the solution to $\sqrt{x} - 1 = 3$ is $x = 16$.

IAAT PRACTICE TEST 5

SECTION 1
Time — 10 minutes
15 Questions

Directions: Read the information given and choose the best answer for each question. Base your answer only on the information given. The time limit for each section is 10 minutes.

1. Today is Monday. What day of the week is it 11 days from today?

    (A) Tuesday
    (B) Wednesday
    (C) Thursday
    (D) Friday

2. Evaluate: $28 + (-35) + 4$

    (J) 11
    (K) 5
    (L) 3
    (M) -3

3. If the perimeter of a rectangular garden is 20 meters and the width is 2 meters, what is the length of the garden in meters?

    (A) 8
    (B) 9
    (C) 10
    (D) 11

4. What is the least common multiple of 9 and 15?

    (J) 27
    (K) 36
    (L) 42
    (M) 45

5. Evaluate: $-3 + 4 \times 5 \div 2$

    (A) 2.5
    (B) 5
    (C) 7
    (D) 9

$$2, 6, 18, 54, \cdots$$

6. What are the next two terms in the sequence above?

    (J) 81 and 162
    (K) 81 and 243
    (L) 162 and 324
    (M) 162 and 486

7. Which of the following represents the greatest integer?

   (A) -7
   (B) $|-6|$
   (C) $\sqrt{25}$
   (D) $(-2)^2$

8. Which of the following is the correct scientific notation of the number 0.000901?

   (J) 9.01×10^3
   (K) 9.01×10^{-3}
   (L) 9.01×10^4
   (M) 9.01×10^{-4}

9. $18 - (-32) + (-17) =$

   (A) 70
   (B) 33
   (C) -33
   (D) -70

10. A coat originally priced at $200 is discounted 15% off. What is the amount of money saved on sale?

    (J) $170
    (K) $150
    (L) $30
    (M) $15

11. Write $\dfrac{1}{5 \times 5 \times 5}$ using a negative exponent.

    (A) 5^3
    (B) 5^{-3}
    (C) $(-5)^{-3}$
    (D) $\dfrac{1}{5^{-3}}$

12. Evaluate: $-5(-4)(6)$

    (J) 120
    (K) 15
    (L) -15
    (M) -120

13. A rectangle has a width of 10 inches. The length is 60% less than that of the width. Find the area of the rectangle.

    (A) 28
    (B) 35
    (C) 40
    (D) 44

$$\frac{2}{3} \times \frac{6}{5} \times \frac{5}{4}$$

14. Simplify the expression above.

    (J) $\frac{12}{5}$

    (K) $\frac{7}{3}$

    (L) $\frac{4}{3}$

    (M) 1

15. Evaluate: $11 \times 65 + 11 \times 35$

    (A) 850
    (B) 900
    (C) 1000
    (D) 1100

STOP

IAAT PRACTICE TEST 5

SECTION 2
Time — 10 minutes
15 Questions

Directions: Read the information given and choose the best answer for each question. Base your answer only on the information given. The time limit for each section is 10 minutes.

Directions: Use the following graph to answer questions 1 – 4.

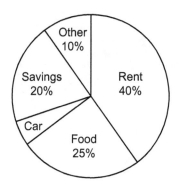

Joshua's Monthly Budget

1. Which of the following is the percent label for car?

   (A) 5%
   (B) 7%
   (C) 10%
   (D) 12%

2. If Joshua's monthly income is $2000, how much does he spend on food each month?

   (J) $250
   (K) $325
   (L) $375
   (M) $500

3. If Joshua's monthly income is $3000, what is his annual budget for rent?

   (A) $1200
   (B) $10000
   (C) $12000
   (D) $14400

4. If Joshua saves $200 a month, what is his monthly income?

   (J) $800
   (K) $900
   (L) $1000
   (M) $1200

Directions: Use the following graph to answer questions 5 – 8.

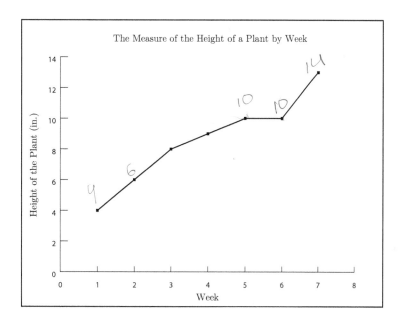

5. How many inches did the plant grow from Week 1 to Week 2?

   (A) 5
   (B) 48
   (C) 3
   (D) 2

6. During which two weeks was there no change in the height of the plant?

   (J) From Week 1 to Week 2
   (K) From Week 3 to Week 4
   (L) From Week 4 to Week 5
   (M) From Week 5 to Week 6

7. How many inches did the plant grow from Week 2 to Week 5?

   (A) 4
   (B) 5
   (C) 6
   (D) 7

8. During which two weeks was there the largest growth rate?

   (J) From Week 6 to Week 7
   (K) From Week 1 to Week 2
   (L) From Week 5 to Week 6
   (M) From Week 1 to Week 5

Directions: Use this information to answer questions 9 – 13.

The ages of a particular family are 23, 25, 42, 5, 3, 9, and 15. The price of the restaurant, listed below, offers various discounts depending on the age of the person.

Under 5	5 – 7	8 – 11	12 or Older
Free	50% Off	25% Off	$20

9. What is the mode age of the family?

   (A) There is not enough information given to answer this question.
   (B) There is no mode.
   (C) 23
   (D) 42

10. What is the median age of the family?

    (J) 3
    (K) 5
    (L) 15
    (M) 23

11. How many family members receive a 50% off discount?

    (A) 0
    (B) 1
    (C) 2
    (D) 3

12. If a 9-year old child receives a 25% off discount, how much money does she save?

    (J) $15
    (K) $12
    (L) $10
    (M) $5

13. How much is the total bill?

    (A) $95
    (B) $105
    (C) $115
    (D) $125

Directions: Use the following venn diagram to answer questions 14 – 15.

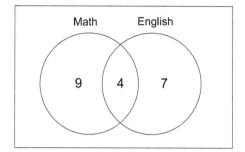

Venn Diagram of Student's Favorite Academic Subject

14. Joshua surveyed a number of students in his class and found out that 9 people favored only math, 7 people favored only English, and 4 people favored both math and English as shown in the venn diagram above. How many students did Joshua survey?

   (J) 28
   (K) 26
   (L) 24
   (M) 20

15. What percent of students favored both math and English as their favorite subject?

   (A) 15%
   (B) 20%
   (C) 25%
   (D) 30%

IAAT PRACTICE TEST 5

SECTION 3
Time — 10 minutes
15 Questions

Directions: Read the information given and choose the best answer for each question. Base your answer only on the information given. The time limit for each section is 10 minutes.

1. A company earns 150% profit of the expenses spent. This relationship is described by the formula $P = 1.5E$, where P is the profit and E is the expenses. If the company earns $75 in profit, how much did it spend in expenses?

    (A) $25
    (B) $50
    (C) $75
    (D) $100

2. What is the slope of the line that passes through the points $(1, 4)$ an $(0, 7)$?

    (J) -3
    (K) -2
    (L) 2
    (M) 3

3. What is the value of y for the equation, $y = -3x + 3$, if $x = 7$?

    (A) -18
    (B) -14
    (C) 18
    (D) 24

4. The number of books sold, y, is three less than double the number of magazines sold, x. If 5 books were sold, how many magazines were sold?

    (J) 3
    (K) 4
    (L) 6
    (M) 7

5. Which table contains only values that satisfy the equation, $y = -\frac{x}{2}$?

(A)
x	1	2	3	4
y	0.5	1	1.5	2

(B)
x	1	2	3	4
y	-2	-4	-6	-8

(C)
x	3	5	7	9
y	1.5	2.5	3.5	4.5

(D)
x	0	1	2	3
y	0	-0.5	-1	-1.5

6. Which of the following table does NOT represent a function?

(J)
x	3	2	3
y	5	7	9

(K)
x	3	5	7
y	2	4	6

(L)
x	1	2	3
y	1	1	1

(M)
x	6	5	4
y	-3	-30	4

7. What is the y-intercept for the line with the equation: $y = x + \frac{1}{2}$.

(A) 0
(B) $\frac{1}{2}$
(C) 1
(D) $\frac{3}{2}$

8. If the perimeter of a square is 16, what is the area of the square?

(J) 64
(K) 48
(L) 36
(M) 16

9. Which of the following equation best represents the following verbal phrase? The number of pencils, p, is six less the number of pens, q.

(A) $p = q - 6$
(B) $q = p - 6$
(C) $p = 6 - q$
(D) $q = 6 - p$

10. Joshua runs away from a starting position 5 meters for a second. The next second, Joshua runs 3 meters towards the starting position. If he continues this pattern, how far away from the starting position would Joshua be after 3 seconds?

(J) 7
(K) 9
(L) 12
(M) 13

11. Which of the following line has a slope that is undefined?

    (A) $x = 5$
    (B) $y = 5$
    (C) $x + y = 5$
    (D) $y = 2x$

12. There are three people: Joshua, Sue, and Mr. Rhee. How many different seating arrangements are possible for the three people?

    (J) 3
    (K) 6
    (L) 9
    (M) 12

13. At a store, two pencils cost $\$x$, and three erasers cost $\$y$. What is the sum of the prices, S, of a pencil and a eraser in dollars?

    (A) $S = 5(x + y)$
    (B) $S = 2x + 3y$
    (C) $S = \frac{x+y}{5}$
    (D) $S = \frac{x}{2} + \frac{y}{3}$

x	y
2	1
4	2
6	3
8	4

14. The table shows four pairs of x and y values. What is true for all values shown in the table above?

    (J) $x = 2y$
    (K) $x = \frac{y}{2}$
    (L) $x = 2y$
    (M) $x = -2y$

	A	B	C	D
Unit Price	$5	$10	$15	$20
Commission	$1	$3	$3	$5

15. Jason works at an electronic store and earns commission based on the items he sells. The table above shows the unit price of each item and the commission he earns per item. Which of the following item earns Jason the highest percentage in commissions?

    (A) Item A
    (B) Item B
    (C) Item C
    (D) Item D

IAAT PRACTICE TEST 5

SECTION 4
Time — 10 minutes
15 Questions

Directions: Read the information given and choose the best answer for each question. Base your answer only on the information given. The time limit for each section is 10 minutes.

1. Simplify: $-3x + 5 + 8x$

    (A) $-10x$
    (B) $-5x + 5$
    (C) $5x + 5$
    (D) $10x$

2. If $x = 3$ and $y = 4$, what is the value of $-2x + 6y + 5$?

    (J) 7
    (K) 11
    (L) 15
    (M) 23

3. Solve for x: $-3x + 1 = -23$

    (A) 3
    (B) 5
    (C) 6
    (D) 8

4. $\angle A$ is a right angle. If the $m\angle A = 15n$, what is the value of n ?

    (J) 8
    (K) 7
    (L) 6
    (M) 5

5. Jason can bike 3 miles in 1 hour. At the same rate, which of the following proportions can be used to find m, the number of miles Jason can bike in 4 hours?

    (A) $\frac{m}{4} = \frac{1}{3}$
    (B) $\frac{m}{3} = \frac{1}{4}$
    (C) $\frac{3}{1} = \frac{m}{4}$
    (D) $\frac{4}{1} = \frac{m}{3}$

6. How many factors does 12 have?

    (J) 6
    (K) 5
    (L) 4
    (M) 3

7. If $2x = 12$ and $x - y = 8$, what is the value of y?

   (A) -2
   (B) 2
   (C) 4
   (D) 6

8. There is a line whose length is 6 feet. If points are placed every 2 feet starting from one end, which of the following represents x, the number of points that lie on the line?

   (J) $x = 2$
   (K) $x = 3$
   (L) $x = 4$
   (M) $x = 5$

   Three more than the product of four and a number, x, is negative thirteen.

9. Solve for x in the verbal phrase above.

   (A) -5
   (B) -4
   (C) -3
   (D) -2

10. Let $d = rt$, where d, r, and t represent distance, rate, and time respectively. What is the distance, in miles, that Mr. Rhee would run if he runs at a rate of 12 miles per hour for 45 minutes?

    (J) 6
    (K) 9
    (L) 12
    (M) 15

11. If $x^3 = 8$, what is the value of x^2?

    (A) 4
    (B) 6
    (C) 8
    (D) 10

12. The sum of the measures of interior angles of a triangle is $180°$. In triangle XYZ, if $m\angle X = 100°$ and $m\angle Y = 55°$, what is $m\angle Z$?

    (J) $15°$
    (K) $25°$
    (L) $35°$
    (M) $45°$

13. If $2x + y = 10$ and $y = 4$, what is the value of $2x$?

    (A) 7
    (B) 6
    (C) 4
    (D) 3

14. The perimeter of a rectangle is 50. If the length is five more than the width, what is the length of the rectangle?

(J) 5
(K) 10
(L) 12
(M) 15

$$\frac{2t^5}{8t^4}$$

15. Write the expression above in simplest form.

(A) $4t^2$
(B) $4t$
(C) $\frac{t}{4}$
(D) $\frac{t}{8}$

STOP

Answers and Solutions
IAAT Practice Test 5 Section 1

Answers

1. D	2. M	3. A	4. M	5. C
6. M	7. B	8. M	9. B	10. L
11. B	12. J	13. C	14. M	15. D

Solutions

1. (D)

    Today is Monday. 1 day from today would be Tuesday, 2 days from today would be Wednesday and so on and so forth. Following this pattern, 7 days from today will be Monday. Thus, the 8th day is Tuesday, 9th day is Wednesday, 10th day is Thursday, and the 11th day is Friday.

2. (M)

    Using the commutative property of addition, add first and then subtract. Since the negative number has a greater value than the positive number, the solution will be negative.

    $$28 + (-35) + 4 = 28 + 4 + (-35) \quad \text{(Use Commutative Property of Addition)}$$
    $$= 32 + (-35)$$
    $$= -3$$

    Therefore, (M) is the correct answer.

3. (A)

    The perimeter of a rectangle is defined as $2\ell + 2w$, where ℓ is the length and w is the width. Since the perimeter and width are given as 20 meters and 2 meters respectively, plug-in the information into the formula and solve for the length, ℓ.

    $$\text{Perimeter} = 2\ell + 2w \quad \text{(Substitute 20 for perimeter and 2 for } w\text{)}$$
    $$20 = 2\ell + 2(2) \quad \text{(Simplify)}$$
    $$20 = 2\ell + 4 \quad \text{(Subtract 4 from each side)}$$
    $$16 = 2\ell \quad \text{(Divide each side by 2)}$$
    $$\ell = 8$$

    Therefore, the length of the garden, ℓ, is 8 meters.

4. (M)

A multiple of a number is a product of that number and a whole number. If two numbers have the same multiples, it is called a common multiple. It is possible to solve for the least common multiple (LCM) by listing out the multiples for each number.

Multiples of 9: 9, 18, 27, 36, **45**, 54 $\cdots$
Multiples of 15: 15, 30, **45**, 60, $\cdots$

Therefore, the least common multiple of 9 and 15 is 45.

5. (C)

When evaluating mathematical expressions, use the order of operations: PEMDAS. This means evaluate expressions in the parenthesis first, then exponents, then multiplication and division from left to right, and finally addition and subtraction from left to right.

$$\begin{aligned}
-3 + 4 \times 5 \div 2 &= -3 + (4 \times 5) \div 2 & \text{(Multiply 4 and 5 first)} \\
&= -3 + 20 \div 2 & \text{(Divide 20 by 2)} \\
&= -3 + 10 & \text{(Add)} \\
&= 7
\end{aligned}$$

Therefore, $-3 + 4 \times 5 \div 2 = 7$.

6. (M)

The sequence suggest a pattern that every term is 3 times the previous term. Thus, the 5th term is $54 \times 3 = 162$, and the the 6th is $162 \times 3 = 486$. Therefore, the next two terms in the sequence are 162 and 486.

7. (B)

Since $|-6| = 6$, $\sqrt{25} = 5$, and $(-2)^2 = 4$, the expression that represents the greatest integer is $|-6|$. Therefore, (B) is the correct answer.

8. (M)

Scientific notation must be written in the form: $c \times 10^n$, where $1 \leq c < 10$ and n is an integer. In general, positive value of n gives larger value than 10 and negative value of n gives smaller value than 1. Since $0.000901 = 9.01 \times 0.0001$, scientific notation of 0.000901 is 9.01×10^{-4}.

9. (B)

Note that $-(-32) = 32$ and $+(-17) = -17$. Thus,

$$18 - (-32) + (-17) = 18 + 32 - 17 = 33$$

Therefore, $18 - (-32) + (-17) = 33$.

10. (L)

Convert 15% into a decimal by moving the decimal point two places to the left: $15\% = 0.15$. Therefore, the amount of money saved on sale is $200 by $0.15 = \$30$.

11. (B)

Recall the properties of exponents: $\frac{1}{a^n} = a^{-n}$

$$\frac{1}{5 \times 5 \times 5} = \frac{1}{5^3} = 5^{-3}$$

Therefore, (B) is the correct answer.

12. (J)

When multiplying two negative numbers together, the product is positive. For example, $-5 \times (-4) = 20$. Therefore, $-5(-4)(6) = 20 \times 6 = 120$.

13. (C)

A rectangle has a width of 10 inches. The length is 60% less than that of the width. This means that the length is equal to 40% of the width. Thus, the length is $10 \times 0.4 = 4$ inches. The area of a rectangle is $\ell \times w$, where ℓ is the length and w is the width. Therefore, the area of the rectangle is $4 \times 10 = 40$.

14. (M)

$$\frac{2}{3} \times \frac{6}{5} \times \frac{5}{4} = \frac{60}{60} = 1$$

Therefore, (M) is the correct answer.

15. (D)

Instead of multiplying numbers immediately, factor out 11 from the expression.

$$11 \times 65 + 11 \times 35 = 11(65 + 35)$$
$$= 11(100)$$
$$= 1100$$

Therefore, (D) is the correct answer.

SOLOMON ACADEMY — TEST 5 SECTION 2

Answers and Solutions
IAAT Practice Test 5 Section 2

Answers

1. A	2. M	3. D	4. L	5. D
6. M	7. A	8. J	9. B	10. L
11. B	12. M	13. B	14. M	15. B

Solutions

1. (A)

   The sum of all percentages must be 100% to represent a whole. Since rent, food, savings, and other add up to $20\% + 10\% + 40\% + 25\% = 95\%$, the percent label for car must be $100\% - 95\% = 5\%$.

2. (M)

   Joshua's monthly income is $2000 and spends 25% on food each month. Since $25\% = 0.25$, the exact amount he spends each month on food is $\$2000 \times 0.25 = \$500$.

3. (D)

   Joshua's monthly income is $3000 and spends 40% on rent each month. Since $40\% = 0.4$, the exact amount he spends each month on rent is $\$3000 \times 0.4 = \$1200$. There are 12 months in one year. In order to obtain the annual budget, multiply a monthly budget, $1200, by 12. Therefore, Joshua's annual budget for rent is $\$1200 \times 12 = \$14400$.

4. (L)

   Joshua saves $200 a month. This is equivalent to 20% of his monthly income. Since 20% of $1000 is $200, Joshua's monthly income is $1000.

5. (D)

   From week 1 to week 2, the plant grew $6 - 4$ or 2 inches. Therefore, (D) is the correct answer.

6. (M)

   From week 5 to week 6, there was no change in the height of the plant. Therefore, (M) is the correct answer.

7. (A)

   From week 2 to week 3, the plant grew 2 inches. From week 3 to week 4, the plant grew 1 inch. Additionally, from week 4 to week 5, the plant grew 1 inch. Therefore, from week 2 to week 5, the plant grew $2 + 1 + 1 = 4$ inches.

8. (J)

In order to quickly determine which week had the largest growth rate, find the line segment which rises most steep. From week 6 to week 7, the plant grew 3 inches. No other week shows this large of an increase. Therefore, (J) is the the correct answer.

9. (B)

The mode of a set is the number that appears the most. Since there is no age that appears most among the family, there is no mode.

10. (L)

In order to determine the median age of the family, it is necessary to first arrange the given numbers from least to greatest: 3, 5, 9, 15, 23, 25, and 42. Since there are 7 people, the middle number is the 4th number, which is 15.

11. (B)

In order to receive a 50% off discount, the age of the family member should be between 5 and 7. Since there is only one 5-year old child in the family, (B) is the correct answer.

12. (M)

25% means 0.25 or $\frac{1}{4}$. Thus, 25% off discount means 0.25 of $20 or $0.25 \times \$20 = \$5$ is subtracted from $20. Therefore, the amount that the 9-year old child saves is $5.

13. (B)

The ages from least to greatest are 3, 5, 9, 15, 23, 25, and 42. The table says that anyone under the age of 5 eats for free. Thus, the child at age 3 eats for free. Children from ages 5-7 eat for 50% off of $20. Thus, the 5-year old child eats for $10. Children from ages 8-11 eat for 25% off of $20. Thus, the 9-year old child eats for $15. Additionally, there are four family members who eat at regular price. The cost of the four family members is $\$20 \times 4 = \$80$. Therefore, the total bill comes out to $\$10 + \$15 + \$80 = \$105$.

14. (M)

The venn diagram displays information about student's favorite academic subject. 9 people favored only math, 7 people favored only English, and 4 people favored both math and English. Therefore, the total number of students that Joshua surveyed is $9 + 7 + 4 = 20$.

15. (B)

A percent, 1%, means 1 out of 100 or $\frac{1}{100}$. Since 4 out of 20 students surveyed favored both math and English as their favorite subject, $\frac{4}{20} = \frac{20}{100} = 20\%$ of the student surveyed favored both subjects. Therefore, (B) is the correct answer.

SOLOMON ACADEMY   TEST 5 SECTION 3

Answers and Solutions
IAAT Practice Test 5 Section 3

Answers

1. B	2. J	3. A	4. K	5. D
6. J	7. B	8. M	9. C	10. J
11. A	12. K	13. D	14. L	15. B

Solutions

1. (B)

   Substitute 75 for the profit, P, in the given relationship $P = 1.5E$ and solve for E.

   $P = 1.5E$ (Substitute 75 for P)
   $75 = 1.5E$ (Divide each side by 1.5)
   $E = 50$

   Therefore, the amount of money that the company spent in expenses is $50.

2. (J)

   The slope determines the steepness and direction of a line. The slope is defined as $\frac{y_2-y_1}{x_2-x_1}$, where (x_1, y_1) and (x_2, y_2) are two points on the line. Since two ordered pairs are given, it is possible to determine the slope.

   $$\frac{y_2 - y_1}{x_2 - x_1} = \frac{7-4}{0-1} = \frac{3}{-1} = -3$$

   Therefore, the slope of the line that passes through the points $(1, 4)$ an $(0, 7)$ is -3.

3. (A)

   Substitute 7 for x in the given equation to find the value of y.

   $$y = -3x + 3 = -3(7) + 3 = -21 + 3 = -18$$

   Therefore, the value of y when $x = 7$ is -18.

4. (K)

Translate the verbal phrase into a mathematic expression. The number of books sold, y, is three less than double the number of magazines sold, x. This can be represented by the equation $y = 2x - 3$. In order to determine the number of magazines sold, substitute 5 for y and solve for x.

$$y = 2x - 3 \quad \text{(Substitute 5 for } y\text{)}$$
$$5 = 2x - 3 \quad \text{(Add 3 to each side)}$$
$$8 = 2x \quad \text{(Divide each side by 2)}$$
$$x = 4$$

Therefore, the number of magazines sold is 4.

5. (D)

x	$y = -\frac{x}{2}$
0	$y = -\frac{0}{2} = 0$
1	$y = -\frac{1}{2} = -0.5$
2	$y = -\frac{2}{2} = -1$
3	$y = -\frac{3}{2} = -1.5$

Since the equation $y = -\frac{x}{2}$ is given, plug in the x values into the equation to determine the y value as shown in the table above. Check the relationship of all x and y values of the table to determine that the table completely satisfies the equation.

6. (J)

A function relates one input to exactly one output. Thus, an input x cannot have more than one value for an output y. In answer choice (J), the x value of 3 repeats and has two different y values: (3,5) and (3,9). Thus, answer choice (J) is not a function. Therefore, (J) is the correct answer.

7. (B)

The slope-intercept form of a line is $y = mx + b$, where m and b are the slope and y-intercept, respectively. The line, $y = x + \frac{1}{2}$, is written in slope-intercept form. Thus, the slope is 1 and y-intercept is $\frac{1}{2}$. Therefore, (B) is the correct answer.

8. (M)

Since the perimeter of a square is 16, the length of the square is $\frac{16}{4} = 4$. Therefore, the area of the square is $4^2 = 16$.

9. (C)

Note that six less x can be expressed as $6 - x$. Whereas, six less than x can be expressed as $x - 6$. Since the number of pencils, p, is six less the number of pens, q, this can be expressed as $p = 6 - q$.

10. (J)

Joshua runs away from a starting position 5 meters for a second. The next second, Joshua runs 3 meters towards the starting position. This means that Joshua is $5 - 3$ or 2 meters away from the starting position after two seconds. The next second, Joshua runs 5 meters further away from the starting position. Therefore, Joshua is $2 + 5 = 7$ meters away from the starting position after 3 seconds.

11. (A)

Slope determines the steepness and direction of a line and is defined as $\frac{y_2 - y_1}{x_2 - x_1}$. A horizontal line has a slope of zero and a vertical line has a slope that is undefined. Since $x = 5$ represents a vertical line, it has a slope that is undefined.

12. (K)

There are three people: Joshua, Sue, and Mr. Rhee. List all possible different seating arrangements for the three people as shown below.

$$\text{Joshua, Sue, Mr. Rhee}$$
$$\text{Joshua, Mr. Rhee, Sue}$$
$$\text{Sue, Joshua, Mr. Rhee}$$
$$\text{Sue, Mr. Rhee, Joshua}$$
$$\text{Mr. Rhee, Joshua, Sue}$$
$$\text{Mr. Rhee, Sue, Joshua}$$

Therefore, there are 6 possible different seating arrangements.

13. (D)

Two pencils cost $\$x$. Thus, one pencil costs $\$\frac{x}{2}$. Three erasers cost $\$y$. Thus, one eraser costs $\$\frac{y}{3}$. Therefore, the sum of the prices, S, of a pencil and a eraser in dollars is $S = \frac{x}{2} + \frac{y}{3}$.

14. (L)

x	$y = \frac{1}{2}x$
-2	$y = \frac{1}{2}(-2) = -1$
-1	$y = \frac{1}{2}(-1) = -\frac{1}{2}$
0	$y = \frac{1}{2}(0) = 0$
1	$y = \frac{1}{2}(1) = \frac{1}{2}$

All ordered pairs in each answer choice must satisfy the equation $y = \frac{1}{2}x$. Thus, if one of the ordered pair does not satisfy the equation, it is not a solution. The only set of ordered pairs that satisfy the equation is answer choice (L) as displayed in the table above. Therefore, (L) is the correct answer.

15. (B)

It is necessary to determine what percentage commission Jason receives per item sold.

	A	B	C	D
Unit Price	$5	$10	$15	$20
Commission	$1	$3	$3	$5
Percent of Commission	$\frac{1}{5} = 20\%$	$\frac{3}{10} = 30\%$	$\frac{3}{15} = 20\%$	$\frac{5}{20} = 25\%$

Therefore, the item at which Jason earns the highest percentage in commission is item B as shown above. Therefore, (B) is the correct answer.

SOLOMON ACADEMY · Distribution or replication of any part of this page is prohibited. · TEST 5 SECTION 4

Answers and Solutions
IAAT Practice Test 5 Section 4

Answers

1. C	2. M	3. D	4. L	5. C
6. J	7. A	8. L	9. B	10. K
11. A	12. K	13. B	14. M	15. C

Solutions

1. (C)

   In order to simplify the expression $-3x + 5 + 8x$, combine like terms. $-3x$ and $8x$ are like terms because each term consists of only a single variable x. Thus, $-3x$ and $8x$ can be combined as $-3x + 8x = 5x$. Since 5 is a constant, $5x$ and 5 are unlike terms and cannot be combined. Therefore, $-3x + 5 + 8x = 5x + 5$.

2. (M)

   Substitute the values given in the question and evaluate the expression. Use orders of operations: PEMDAS.

   $$\begin{aligned} -2x + 6y + 5 &= -2(3) + 6(4) + 5 &&\text{(Substitute 3 for } x \text{ and 4 for } y\text{)} \\ &= -6 + 24 + 5 &&\text{(Simplify)} \\ &= 23 \end{aligned}$$

   Therefore, the value of $-2x + 6y + 5$ when $x = 3$ and $y = 4$ is 23.

3. (D)

   In order to solve for x, use the reverse order of operations, SADMEP, and inverse operations. When dividing two negative numbers, the quotient is positive.

   $$\begin{aligned} -3x + 1 &= -23 &&\text{(Subtract 1 from each side)} \\ -3x &= -24 &&\text{(Divide each side by } -3\text{)} \\ x &= 8 \end{aligned}$$

   Therefore, the solution to $-3x + 1 = -23$ is $x = 8$.

4. (L)

The measure of angle A can be expressed as $m\angle A$. Since angle A is a right angle, $m\angle A = 90°$. Substitute 90 for $m\angle A$ and solve for n.

$$m\angle A = 15n \qquad \text{(Substitute 90 for } m\angle A\text{)}$$
$$90 = 15n \qquad \text{(Divide each side by 15)}$$
$$n = 6$$

Therefore, the value of n is 6.

5. (C)

Proportions state that two ratios, usually expressed as fractions, are the same. Set up a proportion in term of miles and hours.

$$\frac{\text{Number of Miles}}{\text{Number of Hours}} : \quad \frac{3}{1} = \frac{m}{4}$$

Therefore, (C) is the correct answer.

6. (J)

The factors of 12 are 1, 2, 3, 4, 6, and 12. Therefore, the number of factors of 12 is 6.

7. (A)

If $2x = 12$, then $x = 6$. Thus, substitute 6 for x in $x - y = 8$ and solve for y.

$$x - y = 8 \qquad \text{(Substitute 6 for } x\text{)}$$
$$6 - y = 8 \qquad \text{(Subtract 6 from each side)}$$
$$-y = 2 \qquad \text{(Multiply each side by } -1\text{)}$$
$$y = -2$$

Therefore, the value of y is -2.

8. (L)

There is a line that is 6 feet long and points placed every 2 feet as shown figure below.

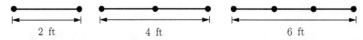

A pattern suggests that if the line is 2 feet long, 4 feet long, and 6 feet long, there are 2 points, 3 points, and 4 points on the line, respectively. Since x represents the number of points that lie on 6-feet line, $x = 4$. Therefore, (L) is the right answer.

9. (B)

Three more than the product of four and a number, x, is negative thirteen can be expressed as $4x + 3 = -13$. In order to solve for x, use the reverse order of operation, SADMEP, and inverse operations. When dividing two negative numbers, the quotient is positive.

$$-4x + 3 = -13 \qquad \text{(Subtract 3 from each side)}$$
$$-4x = -16 \qquad \text{(Divide each side by } -4\text{)}$$
$$x = 4$$

Therefore, the solution to $4x + 3 = -13$ is $x = -4$.

10. (K)

$d = rt$, where d, r, and t represent distance, rate, and time respectively. Before determining the distance that Mr. Rhee would run at a rate of 12 miles per hour for 45 minutes, it is necessary to convert minutes into hours so that the rate and time can have the same unit of time. Since 1 hour is equal to 60 minutes, 45 minutes is $\frac{45}{60}$ or $\frac{3}{4}$ of an hour. Therefore, the distance that Mr. Rhee would run is $d = rt = 12 \times \frac{3}{4} = 9$ miles.

11. (A)

Since $2^3 = 8$, $x = 2$ is the solution to $x^3 = 8$. Therefore, the value of x^2 is $2^2 = 4$.

12. (K)

The sum of the measures of interior angles of a triangle is 180°. Therefore, the measure of angle Z, $m\angle Z$, is $180 - 100 - 55 = 25°$.

13. (B)

Substitute 4 for y in the equation $2x + y = 10$ and solve for $2x$.

$$2x + y = 10 \qquad \text{(Substitute 4 for } y\text{)}$$
$$2x + 4 = 10 \qquad \text{(Subtract 4 from each side)}$$
$$2x = 6$$

Therefore, the value of $2x$ is 6.

14. (M)

The perimeter of a rectangle is 50. The length is five more than the width. Thus, if the width is w, then the length, ℓ, is $\ell = w + 5$.

$$\ell + \ell + w + w = \text{Perimeter} \qquad \text{Substitute 50 for Perimeter}$$
$$w + 5 + w + 5 + w + w = 50 \qquad \text{(Simplify the expression)}$$
$$4w + 10 = 50 \qquad \text{(Subtract 10 from each side)}$$
$$4w = 40 \qquad \text{(Divide each side by 4)}$$
$$w = 10$$

Therefore, the length of the rectangle is $\ell = w + 5 = 10 + 5 = 15$.

15. (C)

Recall the properties of exponents: $\frac{a^m}{a^n} = a^{m-n}$.

$$\frac{2t^5}{8t^4} = \frac{2}{8} \times \frac{t^5}{t^4} \qquad \text{(Since } \frac{t^5}{t^4} = t^{5-4}\text{)}$$
$$= \frac{1}{4} \times t^{5-4} \qquad \text{(Simplify)}$$
$$= \frac{1}{4} \times t$$
$$= \frac{t}{4}$$

Therefore, $\frac{2t^5}{8t^4} = \frac{t}{4}$.

IAAT PRACTICE TEST 6

SECTION 1
Time — 10 minutes
15 Questions

Directions: Read the information given and choose the best answer for each question. Base your answer only on the information given. The time limit for each section is 10 minutes.

1. Jason has 36 puppies. If the ratio of male puppies to female puppies is 1 : 1, how many female puppies are there?

   (A) 18
   (B) 15
   (C) 12
   (D) 9

2. What is the distance on the number line from −15 to 8?

   (J) 21
   (K) 23
   (L) 25
   (M) 27

3. Find the discounted price of a $600 purse that is on sale at 25% off.

   (A) $150
   (B) $300
   (C) $400
   (D) $450

4. There are five light polls on a street equally spaced apart. If the distance from the first poll to the third poll is 18, what is the total distance from the first poll to the last?

   (J) 36
   (K) 45
   (L) 54
   (M) 63

5. There are 8 male dogs and 4 female dogs. If each of the female dogs give birth to 4 puppies, how many dogs are there in total?

   (A) 48
   (B) 36
   (C) 28
   (D) 24

6. Joshua has 120 marbles. Joshua has twice as many marbles as Jason. Jason has 3 times as many marbles as Mr. Rhee. How many marbles does Mr. Rhee have?

   (J) 60
   (K) 40
   (L) 20
   (M) 10

$$2 + 2 \div 2$$

7. Evaluate the following expression above.

    (A) 0
    (B) 1
    (C) 2
    (D) 3

8. Mr. Rhee needs to fill up his car's gas tank. If the cost of gasoline is \$3 per gallon, how many gallons can he buy with \$45?

    (J) 12
    (K) 13
    (L) 14
    (M) 15

$$\frac{3}{5} + \frac{2}{5} - \frac{4}{5}$$

9. Evaluate the following expression above.

    (A) $-\frac{1}{5}$
    (B) $\frac{1}{5}$
    (C) $\frac{2}{5}$
    (D) $\frac{1}{2}$

10. You toss a coin 3 times. How many different outcomes are possible?

    (J) 4
    (K) 6
    (L) 8
    (M) 12

11. Evaluate 3^3.

    (A) $\frac{1}{27}$
    (B) $\frac{1}{9}$
    (C) 9
    (D) 27

12. Jason has 14 red marbles, 4 white marbles, and 2 blue marbles in a bag. If Jason randomly chooses a marble from the bag, what is the probability that the marble is red?

    (J) .7%
    (K) 1.4%
    (L) 0.7
    (M) 0.8

$$3(-2)(4)(-5)$$

13. Evaluate the following expression above.

    (A) −120
    (B) −100
    (C) 46
    (D) 120

14. A lab instructor finds that a germ population count is 10 at 11:00am. If the germ population triples in size every 20 minutes, what is the best estimate of the population at 12:20pm?

    (J) 90
    (K) 270
    (L) 810
    (M) 2430

15. Joshua takes 3 math tests. The average score of his first 2 tests is 85 points. Joshua scored 70 on his third test. What is Joshua's average test score for all 3 tests?

    (A) 70
    (B) 78
    (C) 80
    (D) 82

STOP

IAAT PRACTICE TEST 6

SECTION 2
Time — 10 minutes
15 Questions

Directions: Read the information given and choose the best answer for each question. Base your answer only on the information given. The time limit for each section is 10 minutes.

Directions: Use the following graph to answer questions 1 – 4.

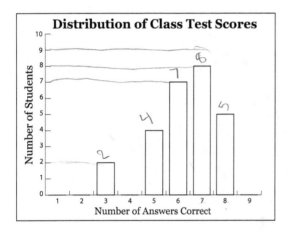

1. How many students are there in the class who took the test?

    (A) 8
    (B) 20
    (C) 24
    (D) 26

2. What is the range of the number of correct answers received by students?

    (J) 8
    (K) 7
    (L) 6
    (M) 5

3. If a student is chosen at random, what is the probability that the student received 7 scores correct?

    (A) $\frac{9}{26}$
    (B) $\frac{8}{26}$
    (C) $\frac{7}{26}$
    (D) $\frac{4}{26}$

4. If half of the students passed, what is the least amount of answers correct needed to pass?

    (J) 7
    (K) 6
    (L) 5
    (M) 4

Directions: Use the following graph to answer questions 5 – 7.

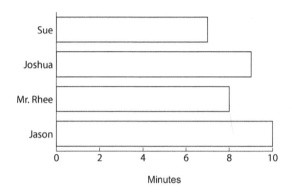

5. Who had the fastest time for running the mile?

   (A) Sue
   (B) Joshua
   (C) Mr. Rhee
   (D) Jason

6. What is the mile run time of Mr. Rhee?

   (J) 7 minutes
   (K) 8 minutes
   (L) 9 minutes
   (M) 10 minutes

7. Assuming that Jason runs at the same rate, how long will it take him to run 3 miles?

   (A) 21 minutes
   (B) 24 minutes
   (C) 27 minutes
   (D) 30 minutes

Directions: Use the following graph to answer questions 8 – 10.

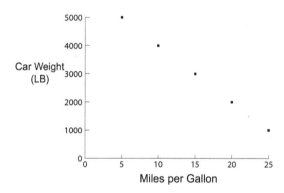

8. A car that weighs 3000 pounds will have what gas mileage?

   (J) 5
   (K) 10
   (L) 15
   (M) 20

9. According to the scatter plot, which of the following statement is true?

   (A) As the weight of the car increases, the gas mileage increases.
   (B) As the weight of the car increases, the gas mileage decreases.
   (C) As the weight of the car decreases, the gas mileage decreases.
   (D) The gas mileage depends on the type of car rather than weight.

10. If your car gets 23 miles per gallon, what is the approximate weight, in pounds, of the car?

    (J) 4500
    (K) 3500
    (L) 2500
    (M) 1500

Pictures	Time
5	20 Minutes

11. If printing at the same rate shown above, how long will it take a photo printer to print 12 pictures.

    (A) 24
    (B) 48
    (C) 54
    (D) 60

Directions: Use set A shown below to answer questions 12 – 14.

$$A = \{8, 14, 11, 5, 7\}$$

12. If all numbers in set A are arranged from least to greatest, which of the following number is the second smallest?

    (J) 14
    (K) 11
    (L) 8
    (M) 7

13. What is the mean of set A?

    (A) 9
    (B) 10
    (C) 11
    (D) 12

14. If 5 is removed from set A, what is the new mean of the set?

    (J) 11
    (K) 10
    (L) 9
    (M) 8

Drink	Number of Students
Water	7
Milk	4
Juice	8
Soda	6

15. The table above shows the favorite drink of students. What percent of students surveyed chose juice as their favorite drink?

   (A) 8%
   (B) 25%
   (C) 30%
   (D) 32%

STOP

IAAT PRACTICE TEST 6

SECTION 3
Time — 10 minutes
15 Questions

Directions: Read the information given and choose the best answer for each question. Base your answer only on the information given. The time limit for each section is 10 minutes.

1. Which of the following scatter plot does NOT represent a function?

(A)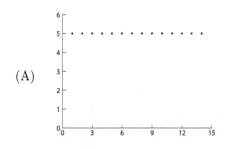

(B)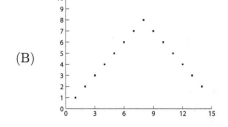

(C)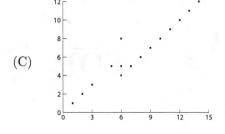

(D)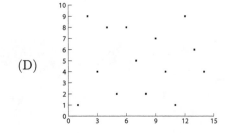

2. A college group is hosting a heritage night. The cost of venue is $2000 plus an additional $6 per guest. Which of the following equation represents the total amount of money, m, that the event will cost the college group if g represents the number of guests attending?

(J) $m = 2000 + 6g$

(K) $m = 6g - 2000$

(L) $m = (2000 + 6)g$

(M) $m = 2000g + 6$

x	1	2	3	4
y	6	10	14	18

3. The table shows four pairs of x and y values. What is true for all values shown in the table above?

(A) $y = x + 5$

(B) $y = 3x + 1$

(C) $y = 6x - 2$

(D) $y = 4x + 2$

162

4, 5, 7, 10, 14, ···

4. What is the next term in the sequence above?

   (J) 18
   (K) 19
   (L) 20
   (M) 21

Input	Output
1	−3
2	−2
3	−1
k	

5. Observe the numbers in the two columns in order to determine which of the following belongs in the empty cell.

   (A) $2x - 8$
   (B) $8 - 2k$
   (C) $k - 4$
   (D) $4 - k$

6. Find the x-intercept of the line that represents the equation, $y = 5x - 15$.

   (J) 1
   (K) 3
   (L) 5
   (M) 7

7. Which of the following equation best represents the following verbal relationship? One-third of the number of cats, c, is ten more than two times the number of dogs, d.

   (A) $3c = 2d + 10$
   (B) $3c = 2c - 10$
   (C) $\frac{c}{3} = 2d + 10$
   (D) $\frac{c}{3} = 2d - 10$

8. Which of the following could be a value of x that satisfies $\frac{1}{2} < x < \frac{2}{3}$?

   (J) 0.4
   (K) 0.5
   (L) 0.6
   (M) 0.7

9. Which of the following table best represents the following verbal relationship? The number of necklaces sold, y, is twelve less than four times the number of bracelets sold, x.

(A)
x	1	2	3	4
y	12	24	36	48

(B)
x	3	4	5	6
y	0	4	8	12

(C)
x	1	2	3	4
y	−8	0	8	16

(D)
x	3	5	7	9
y	0	8	16	20

10. What is the slope for the line with the equation $4x - 2y = 3$?

   (J) $\frac{1}{4}$
   (K) $\frac{1}{2}$
   (L) 2
   (M) 4

11. When $x = 5$, what is the value of y for the equation, $y = 6x + 5$?

   (A) 30
   (B) 35
   (C) 40
   (D) 45

12. The postage cost for a package is $0.45 for the first ounce and $0.23 for each additional ounce. Which of the following is an expression that best represents the cost of a package that weighs 10 ounces in dollars?

   (J) $0.45 + 0.23(10)$
   (K) $0.45(10) + 0.23$
   (L) $0.45(9) + 0.23$
   (M) $0.45 + 0.23(9)$

13. Joshua is three times as old as Jason. What is the ratio of Joshua's to Jason's ages?

   (A) $3:1$
   (B) $2:1$
   (C) $1:1$
   (D) $1:3$

14. Joshua, J, has four more points than Alex, A. Which equation represents the number of points that Alex has?

   (J) $A = J - 4$
   (K) $A = J + 4$
   (L) $A = 4J$
   (M) $A = \frac{J}{4}$

x	y
1	2
2	2
3	2
⋮	⋮
6	

15. If the table above shows a set of ordered pairs on a line, what is the value of y when $x = 6$?

   (A) 2
   (B) 3
   (C) 4
   (D) 5

STOP

IAAT PRACTICE TEST 6

SECTION 4
Time — 10 minutes
15 Questions

Directions: Read the information given and choose the best answer for each question. Base your answer only on the information given. The time limit for each section is 10 minutes.

1. If $a \star b = b - a$, what is $4 \star 5$?

   (A) 1
   (B) 2
   (C) 3
   (D) 4

2. Express the following verbal phrase algebraically: A number, x, is decreased by 12.

   (J) $\frac{x}{12}$
   (K) $12x$
   (L) $12 - x$
   (M) $x - 12$

3. If $x = 8$ and $y = -12$, simplify $2x + 2y + 4z$.

   (A) $4z + 8$
   (B) $4z - 8$
   (C) $8 - 4z$
   (D) $8z - 4$

4. Two angles are considered supplementary if the sum of the measures of the two angles is 180°. $\angle A$ and $\angle B$ are supplementary angles. If the $m\angle A = 115°$, what is the $m\angle B$?

   (J) 35°
   (K) 55°
   (L) 65°
   (M) 85°

5. If $x = 6$, what is the value of $\frac{x^2}{2}$?

   (A) 18
   (B) 12
   (C) 9
   (D) 6

6. If $\frac{x}{2} = 1$, what is the value of $3x$?

   (J) 8
   (K) 6
   (L) 4
   (M) 2

7. Solve for x: $-3x < 15$

   (A) $x < 5$
   (B) $x < -5$
   (C) $x > -5$
   (D) $x > 5$

8. Which of the following expression is equal to $2a \times 2a$?

   (J) $4a$
   (K) $2a^2$
   (L) $(2a)^2$
   (M) $(4a)^2$

9. $P = S - E$ is an equation that relates earned profit, P, to the difference of sales, S, and expenses, E. Sue's monthly expense is always $1250. If Sue wants to make a profit of $3400 next month, how many dollars must her shop earn in sales?

   (A) $4650
   (B) $2150
   (C) $1250
   (D) $900

$$-8x \times -3x$$

10. Simplify the following expression above.

    (J) $24x^2$
    (K) $-24x^2$
    (L) $11x$
    (M) $-11x$

11. Solve the equation: $\frac{1}{4}(2 - x)$ if $x = \frac{1}{2}$.

    (A) $\frac{5}{8}$
    (B) $\frac{1}{2}$
    (C) $\frac{3}{8}$
    (D) $\frac{1}{4}$

12. The lengths of the sides of a triangle are x, y, and 3. If the perimeter of the triangle is 11, which of the following would be the equation that describes the perimeter of the triangle?

    (J) $x + y = 11$
    (K) $x + y = 11 + 3$
    (L) $x + y = 11 \times 3$
    (M) $x + y + 3 = 11$

13. Express the following expression verbally: $2x + 4y$.

   (A) The sum of two times a number, x, and another number, y.

   (B) The sum of two times a number, x, and four times another number, y.

   (C) The product of two times a number, x, and four times another number, y.

   (D) Two times the sum of a number, x, and another number, y, plus four.

14. Solve for x: $\frac{1}{2}(x-1) = 7$.

   (J) 15
   (K) 16
   (L) 17
   (M) 18

15. The volume of a cone is defined as $\frac{1}{3}\pi r^2 h$, where r and h represent the radius and height of the cone, respectively. What is the volume of a cone if the radius is 3 and the height is 2?

   (A) 3π
   (B) 6π
   (C) 12π
   (D) 18π

STOP

SOLOMON ACADEMY — TEST 6 SECTION 1

Answers and Solutions
IAAT Practice Test 6 Section 1

Answers

1. A	2. K	3. D	4. J	5. C
6. L	7. D	8. M	9. B	10. L
11. D	12. L	13. D	14. L	15. C

Solutions

1. (A)

   There are 36 puppies in total. Since the ratio of male puppies to female puppies is 1:1, the number of female puppies is half the total number of puppies. Therefore, the number of female puppies is $36 \times \frac{1}{2} = 18$.

2. (K)

   Subtracting a negative number means to add. Therefore, the distance on the number line from -15 to 8 is $8 - (-15) = 23$.

3. (D)

   The $600 purse is on sale at 25% off. In order words, the new cost of the purse is $100\% - 25\% = 75\%$ of the original price. Convert 75% into a decimal by dividing by 100 or moving the decimal point two places to the left. Thus, $75\% = 0.75$. Therefore, the price of the purse after the discount is $\$600 \times 0.75 = \$450$.

4. (J)

   There are 2 equal intervals of length in between the first poll to the third poll. Since the distance from the first poll to the third poll is 18, the length of each interval is $\frac{18}{2} = 9$. If there are five light polls on a street, there are 4 equal intervals of length in between the light polls. Therefore, the distance from the first poll to the fifth poll is $9 \times 4 = 36$.

5. (C)

   There are 8 male dogs and 4 female dogs. Each of the 4 female dogs give birth to 4 puppies. Thus, there are $4 \times 4 = 16$ puppies. Therefore, the total number dogs is $8 + 4 + 16 = 28$.

6. (L)

   Joshua has twice as many marbles as Jason, which means that Jason has half as many marbles as Joshua. Since Joshua has 120 marbles, Jason has 60 marbles. Since Jason has 3 times as many marbles as Mr. Rhee, Mr. Rhee has $\frac{1}{3}$ of the marbles that Jason has. Therefore, Mr. Rhee has $\frac{1}{3} \times 60 = 20$ marbles.

7. (D)

When evaluating mathematical expressions, use the order of operations: PEMDAS. This means to solve expressions in the parenthesis first, then exponents, then multiplication and division from left to right, and finally addition and subtraction from left to right.

$$2 + 2 \div 2 = 2 + (2 \div 2) \qquad \text{(Divide before Adding)}$$
$$= 2 + 1$$
$$= 3$$

Therefore, (D) is the correct answer.

8. (M)

The cost of gasoline is \$3 per gallon. In order to find how many gallons Mr. Rhee can buy with \$45, use division: $45 \div 3 = 15$. Therefore, Mr. Rhee can buy 15 gallons.

9. (B)

In order to add or subtract fractions, they must have a common denominator. Since all of the fractions in the expression have the same denominator, proceed to add and subtract from left to right.

$$\frac{3}{5} + \frac{2}{5} - \frac{4}{5} = \frac{3 + 2 - 4}{5} = \frac{1}{5}$$

Therefore, (B) is the correct answer.

10. (L)

$$\begin{array}{llll} H, H, H & H, T, H & H, H, T & H, T, T \\ T, T, T & T, H, T & T, T, H & T, H, H \end{array}$$

Since each flip of the coin has two possible outcomes, head (H) or tail (T), you have $2 \times 2 \times 2 = 8$ different possible outcomes if you flip a coin 3 times. The 8 different possible outcomes are shown above.

11. (D)

The exponent of a power describes how many times the base number is multiplied by itself.

$$3^3 = 3 \times 3 \times 3 = 9 \times 3 = 27$$

Therefore, $3^3 = 27$.

12. (L)

Probability is the likelihood of an event to occur. Since there are 14 red marbles, 4 white marbles, and 2 blue marbles, there is a total of $14 + 4 + 2 = 20$ marbles. Out of the 20 marbles, there are 14 red marbles. Thus,

$$\text{Probability} = \frac{\text{Number of Red Marbles}}{\text{Number of Total Marbles}} = \frac{14}{20} = \frac{7}{10}$$

Therefore, the probability of choosing a red marble from the bag is $\frac{7}{10}$ or 0.7.

13. (D)

When numbers that are directly next to each other are separated by nothing but a parenthesis, it means to multiply. Therefore, $3(-2)(4)(-5) = 3 \times -2 \times 4 \times -5 = 120$.

14. (L)

A lab instructor finds that a germ population count is 10 at 11:00am and triples in size every 20 minutes. In order to find the germ population count at 12:20pm, it is necessary to find out the number of 20 minute intervals between 11:00am and 12:20pm. There are 4 intervals between 11:00am and 12:20pm: 11:00-11:20, 11:20-11:40, 11:40-12:00, and 12:00-12:20. Since the germ count triples every 20 minutes, the germ count is $3^4 = 3 \times 3 \times 3 \times 3 = 81$ times the original value. Therefore, the germ population at 12:20pm is $10 \times 81 = 810$.

15. (C)

The average, or mean, is the total sum of elements divided by the number of elements. It may be necessary to determine the sum when an average is given: Sum = Average × Number. Since the average score of Joshua's first two tests is 85, he scored a total of $85 \times 2 = 170$ for the two tests. Since the score of his third test is 70, the total score for the three tests is $170 + 70 = 240$. Thus, in order to determine the average score for all 3 tests, divide the total score of all three tests by 3. Therefore, the average score of the three tests is $\frac{240}{3} = 80$.

SOLOMON ACADEMY — TEST 6 SECTION 2

Answers and Solutions
IAAT Practice Test 6 Section 2

Answers

1. D	2. M	3. B	4. J	5. A
6. K	7. D	8. L	9. B	10. M
11. B	12. M	13. A	14. K	15. D

Solutions

1. (D)

   The number of students can be observed by looking at the y-axis. There were 2 students who scored 3, 4 students who scored 5, 7 students who scored 6, 8 students who scored 7, and 5 students who scored 8. Therefore, the total number of students in the class who took the test is $2 + 4 + 7 + 8 + 5 = 26$.

2. (M)

   The range of a data set is the difference between the highest and lowest values. The highest number of correct answers is 8 and the lowest number of correct answers is 3. Therefore, the range of the number of correct answers received by students is $8 - 3 = 5$.

3. (B)

   Probability means the number of favorable outcomes divided by the total possible number of outcomes. Out of the 26 total students, 8 people scored 7, Therefore, the probability that a randomly chosen student received 7 scores correct is $\frac{8}{26}$.

4. (J)

   Half of the students passed the class. Since there were 26 students who took the test, $\frac{26}{2}$ or 13 students passed. Observe the graph to determine what the least amount of correct answers is needed to pass. Since 13 people in total scored either a 7 or 8, the minimum number of correct answers needed to pass is 7.

5. (A)

   Be careful in how you interpret data given in the form of a graph. A common mistake is to say that Jason finished the fastest because his bar is the longest. However, since the bar represents minutes and the fastest person takes the least amount of time, Sue had the fastest time for running the mile because her bar is the shortest. Therefore, (A) is the correct answer.

6. (K)

   The mile run time of Mr. Rhee is 8 minutes as shown in the graph.

7. (D)

Jason ran 1 mile in 10 minutes. At the same rate, it would take Jason $10 \times 3 = 30$ minutes to run 3 miles.

8. (L)

The ordered pair $(15, 3000)$ in the graph indicates that a gas mileage is 15 when a car weighs 3000 pounds. Therefore, (L) is the correct answer.

9. (B)

The graph shows a pattern such that as the car weight increases, the gas mileage decreases. Therefore, (B) is the correct answer.

10. (M)

In order to estimate the approximate weight, in pounds, of the car that gets 23 miles per gallon, observe the graph. A car weighing 2000 pounds gets 20 miles per gallon as where a car weighing 1000 pounds gets 25 miles per gallon. Since 23 miles per gallon is in between 20 and 25, the car's weight is in between 1000 and 2000 pounds. Therefore, the car must weigh approximately 1500 pounds.

11. (B)

If a photo printer prints 5 pictures in 20 minutes, it prints one picture every $\frac{20}{5} = 4$ minutes. Therefore, it will take $12 \times 4 = 48$ minutes to print 12 pictures.

12. (M)

If all numbers in set A are arranged from least to greatest, $A = \{5, 7, 8, 11, 14\}$. Therefore, the second smallest number in set A is 7.

13. (A)

The mean, or the average, is defined as the sum of all elements divided by the number of elements. Since $A = \{8, 14, 11, 5, 7\}$,

$$\text{Mean} = \frac{\text{Sum of all elements}}{\text{Number of elements}}$$
$$= \frac{8 + 14 + 11 + 5 + 7}{5}$$
$$= \frac{45}{5}$$
$$= 9$$

Therefore, the mean of set A is 9.

SOLOMON ACADEMY — Distribution or replication of any part of this page is prohibited. — TEST 6 SECTION 2

14. (K)

If 5 is removed from set A, set A becomes $A = \{8, 14, 11, 7\}$. Thus,

$$\text{New mean} = \frac{8 + 14 + 11 + 7}{4}$$
$$= \frac{40}{4}$$
$$= 10$$

Therefore, the new mean of set A is 10.

15. (D)

Among the students surveyed, 8 students chose juice as their favorite drink. Since $7+4+8+6 = 25$ people were surveyed in all, $\frac{8}{25}$ of students surveyed chose juice as their favorite drink. In order to find the percentage, multiply both the numerator and denominator of $\frac{8}{25}$ by 4.

$$\text{Percent} = \frac{8}{25} = \frac{8 \times 4}{25 \times 4} = \frac{32}{100} = 32\%$$

Therefore, 32% of the students surveyed chose juice as their favorite drink.

SOLOMON ACADEMY · TEST 6 SECTION 3

Answers and Solutions
IAAT Practice Test 6 Section 3

Answers

1. C	2. J	3. D	4. K	5. C
6. K	7. C	8. L	9. B	10. L
11. B	12. M	13. A	14. J	15. A

Solutions

1. (C)

   A function relates an input to an output. Thus, an input x cannot have more than one value for an output y. When a scatter plot or line graph is given, it is possible to determine whether or not it is a function by the vertical line test. If a vertical line can be drawn at any location and crosses through multiple points, it fails the vertical line test and is NOT a function. This is evident in answer choice (C), as the x-value of 6 has multiple y-values at 4, 5, and 8. Thus, answer choice (C) is NOT a function. Therefore, (C) is the correct answer.

2. (J)

   A college group is hosting a heritage night and the cost of venue is \$2000. This is the starting cost of the event and is the y-intercept. Since the event costs the college group \$6 per guest, 6 is the slope. Therefore, the total amount of money, m, that the event will cost the college group can be written as $m = 2000 + 6g$, where g is the number of guests attending. Therefore, (J) is the correct answer.

3. (D)

x	$y = 4x + 2$
1	$y = 4(1) + 2 = 6$
2	$y = 4(2) + 2 = 10$
3	$y = 4(3) + 2 = 14$
4	$y = 4(3) + 2 = 18$

   Plug in the x values into the equations listed as answer choices to determine which would yield the desired y values. The equation $y = 4x + 2$ is the only equation that satisfies the four ordered pairs $(1, 6)$, $(2, 10)$, $(3, 14)$, and $(4, 18)$. Therefore, (D) is the correct answer.

4. (K)

$$4, \xrightarrow{+1} 5, \xrightarrow{+2} 7, \xrightarrow{+3} 10, \xrightarrow{+4} 14, \xrightarrow{+5} 19$$

The sequence suggests a pattern that each term increases by 1, 2, 3, 4, and 5 as shown above. Therefore, the next term in the sequence is $14 + 5 = 19$.

5. (C)

Input	Output
1	$1 - 4 = -3$
2	$2 - 4 = -2$
3	$3 - 4 = -1$
k	$k - 4$

When the values of input are given, the values of output is 4 less than the values of input. Therefore, when the value of input is k, the value of output is $k - 4$.

6. (K)

The x-intercept is the point at which the line crosses the x-axis. In order to find the x-intercept, set y to zero and solve for x in the equation.

$$\begin{aligned} y &= 5x - 15 & &\text{(Set } y \text{ to zero)} \\ 0 &= 5x - 15 & &\text{(Add 15 to each side)} \\ 15 &= 5x & &\text{(Divide each side by 5)} \\ x &= 3 \end{aligned}$$

Therefore, the x-intercept of the line $y = 5x - 15$ is 3.

7. (C)

Verbal Phrase	Expression
Two times the number of dogs, d	$2d$
Ten more than two times the number of dogs, d	$2d + 10$

Since one-third the number of cats, $\frac{c}{3}$, is ten more than two times the number of dogs, d, the equation that represents the verbal relationship is $\frac{c}{3} = 2d + 10$.

8. (L)

Since $\frac{1}{2} = 0.5$ and $\frac{2}{3} = 0.66\cdots$, the value of x that satisfies $\frac{1}{2} < x < \frac{2}{3}$ could be 0.6. Therefore, (L) is the correct answer.

9. (B)

x	$y = 4x - 12$
3	$y = 4(3) - 12 = 0$
4	$y = 4(4) - 12 = 4$
5	$y = 4(5) - 12 = 8$
6	$y = 4(6) - 12 = 12$

The number of necklaces sold, y, is twelve less than four times the number of bracelets sold, x, can be written as $y = 4x - 12$. All ordered pairs in each answer choice must satisfy the equation $y = 4x - 12$. The only set of ordered pairs that satisfy the equation is answer choice (B) as displayed in the table above. Therefore, (B) is the correct answer.

10. (L)

Rewrite the equation so that y is a function of x, or y in terms of x.

$$4x - 2y = 3 \qquad \text{(Subtract each side by } 4x\text{)}$$
$$-2y = -4x + 3 \qquad \text{(Divide each side by } -2\text{)}$$
$$y = 2x - \frac{3}{2}$$

The equation $y = 2x - \frac{3}{2}$ is written in slope-intercept form, $y = mx + b$, where m and b represent the slope and y-intercept respectively. Therefore, the slope of $y = 2x - \frac{3}{2}$ is 2.

11. (B)

Substitute 5 for x and solve for y.

$$y = 6x + 5 \qquad \text{(Substitute 5 for } x\text{)}$$
$$= 6(5) + 5 \qquad \text{(Simplify)}$$
$$= 35$$

Therefore, the value of y when $x = 5$ is 35.

12. (M)

The postage cost for a package is $0.45 for the first ounce and $0.23 for each additional ounce. Thus, if a package weighs 10 ounces, the first ounce costs 45 cents and the remaining $10 - 1 = 9$ ounces cost 23 cents each, which can be expressed as $0.45 + 0.23(9)$. Therefore, (M) is the correct answer.

13. (A)

A ratio is a fraction that compares two quantities measured in the same units The ratio of a to b can be written as $a : b$ or $\frac{a}{b}$. Since Joshua is three times as old as Jason, the ratio of Joshua's to Jason's age is $3 : 1$.

14. (J)

Joshua, J, has four more points than Alex, A. Thus, $J = A+4$. In order to determine the equation that represents the number of points Alex has, A, subtract 4 from each side of $J = A + 4$.

$$J = A + 4 \qquad \text{(Subtract 4 from each side)}$$
$$A = J - 4$$

Therefore, (J) is the correct answer.

15. (A)

x	y
1	2
2	2
3	2
4	2
5	2
6	2

The table above represents a set of ordered pairs on a horizontal line $y = 2$ because the values of y remains 2 when the values of x changes. Therefore, when $x = 6$, the value of y is 2.

SOLOMON ACADEMY — TEST 6 SECTION 4

Answers and Solutions
IAAT Practice Test 6 Section 4

Answers

1. A	2. M	3. B	4. L	5. A
6. K	7. C	8. L	9. A	10. J
11. C	12. M	13. B	14. J	15. B

Solutions

1. (A)

   A function $a\bigstar b$ is defined as $b - a$. In order to determine $4\bigstar 5$, substitute 4 for a and 5 for b in the expression $b - a$. Therefore, $4\bigstar 5 = 5 - 4 = 1$.

2. (M)

   A number, x, is decreased by 12 can be expressed as $x - 12$. Therefore, (M) is the correct answer.

3. (B)

   After substituting 8 for x and -12 for y in $2x + 2y + 4z$, combine like terms and simplify.

   $$2x + 2y + 4z = 2(8) + 2(-12) + 4z = 16 - 24 + 4z = 4z - 8$$

   Since $4z$ and -8 are unlike terms, they cannot be combined. Therefore, if $x = 8$ and $y = -12$, the expression $2x + 2y + 4z$ simplifies to $4z - 8$.

4. (L)

   Two angles are considered supplementary if the sum of the measures of the two angles is $180°$. $\angle A$ and $\angle B$ are supplementary. Thus, $m\angle A + m\angle B = 180°$. Since the $m\angle A = 115°$, the $m\angle B = 180 - 115 = 65°$.

5. (A)

   Substitute 6 for x and evaluate the expression.

   $$\frac{x^2}{2} = \frac{6^2}{2} = \frac{36}{2} = 18$$

   Therefore, the value of $\frac{x^2}{2} = 18$.

6. (K)

   Since $\frac{x}{2} = 1$, $x = 2$. Therefore, the value of $3x$ is $3(2) = 6$.

7. (C)

It is important to note that when multiplying or dividing an inequality with a negative number, the inequality symbol is reversed.

$$-3x < 15 \quad \text{(Divide each side by } -3 \text{ and reverse the inequality symbol)}$$
$$x > -5$$

Therefore, the solution to $-3x < 15$ is $x > -5$.

8. (L)

x^2 means x is multiplied by itself twice. Since $2a$ is multiplied by itself twice, $2a \times 2a$ can be expressed as $(2a)^2$. Therefore, (L) is the correct answer.

9. (A)

$P = S - E$ is an equation that relates earned profit, P, to the difference of sales, S, and expenses, E. Since Sue's monthly expensive is always \$1250, substitute 1250 for E in the equation. Furthermore, Sue wants to earn a profit of \$3400. Thus, substitute 3400 for P. Then, the equation can be written as \$3400 = S − \$1250. Therefore, the amount of sales, S, is S = \$3400 + \$1250 = \$4650.

10. (J)

Recall the property of exponents: $a^m \times a^n = a^{m+n}$.

$$-8x \times -3x = -8 \times -3 \times x \times x$$
$$= 24 \times x^2$$
$$= 24x^2$$

Therefore, (J) is the correct answer.

11. (C)

Substitute $\frac{1}{2}$ for x in the expression $\frac{1}{4}(2-x)$ and evaluate.

$$\frac{1}{4}\left(2 - \frac{1}{2}\right) = \frac{1}{4}\left(\frac{4}{2} - \frac{1}{2}\right) = \frac{1}{4}\left(\frac{3}{2}\right) = \frac{3}{8} \tag{2}$$

Therefore, (C) is the correct answer.

12. (M)

The perimeter a triangle is the distance around the triangle. If the lengths of the sides of a triangle are x, y, and 3, the perimeter of the triangle can be expressed as $x + y + 3$. Since the perimeter of the triangle is 11, $x + y + 3 = 11$.

13. (B)

Expression	Verbal Phrase
$2x$	Two times a number, x
$4y$	Four times another number, y
$2x + 4y$	The sum of two times a number, x, and four times another number, y

Therefore, (B) is the correct answer.

14. (J)

$$\frac{1}{2}(x-1) = 7 \quad \text{(Multiply each side by 2)}$$
$$x - 1 = 14 \quad \text{(Add 1 to each side)}$$
$$x = 15$$

Therefore, the solution to $\frac{1}{2}(x-1) = 7$ is $x = 15$.

15. (B)

The volume of a cone is defined as $\frac{1}{3}\pi r^2 h$, where r and h represent the radius and height of the cone, respectively. The radius and height of the cone are given as 3 and 2 respectively. Substitute the given values into the equation and solve for the volume.

$$\text{Volume} = \frac{1}{3}\pi r^2 h \quad \text{(Substitute 3 for } r \text{ and 2 for } h\text{)}$$
$$= \frac{1}{3}\pi (3)^2 (2) \quad \text{(Since } (3)^2(2) = 18\text{)}$$
$$= \frac{1}{3}\pi (18) \quad \text{(Simplify)}$$
$$= 6\pi$$

Therefore, the volume of a cone with the radius and height of 3 and 2, respectively, is 6π.

SOLOMON ACADEMY — Distribution or replication of any part of this page is prohibited. — TEST 7 SECTION 1

IAAT PRACTICE TEST 7

SECTION 1
Time — 10 minutes
15 Questions

Directions: Read the information given and choose the best answer for each question. Base your answer only on the information given. The time limit for each section is 10 minutes.

1. What is the value of $4\frac{2}{3} \div \frac{1}{3}$?

    (A) $\frac{14}{9}$
    (B) $4\frac{2}{9}$
    (C) 8
    (D) 14

2. Simplify the following expression: $15 \div 0.5$

    (J) 30
    (K) 25
    (L) 12
    (M) $7\frac{1}{2}$

3. If a scanner can scan 1 page of a document every 9 seconds, how many pages can the scanner scan in 3 minutes?

    (A) 8
    (B) 16
    (C) 20
    (D) 24

4. Evaluate: $3 \times 10^3 \times 5 \times 10^2$

    (J) 1.5×10^4
    (K) 15×10^5
    (L) 1.5×10^6
    (M) 15×10^6

5. Joshua spent $23 in January, $25 in February, $55 in March, and $57 in April. What is the total amount Joshua spent during the four months?

    (A) $125
    (B) $130
    (C) $160
    (D) $175

6. Evaluate the following expression: $\frac{6}{21} - \frac{23}{84}$

    (J) $\frac{1}{84}$
    (K) $\frac{1}{42}$
    (L) $\frac{1}{28}$
    (M) $\frac{1}{21}$

7. The original cost of a computer is $1000. On discount, Jason paid $750 for the computer. What percentage off was the computer on sale?

   (A) 20%
   (B) 25%
   (C) 30%
   (D) 35%

8. Joshua can walk 5 blocks in 4 minutes. If each block is 20 feet long, how many feet will he walk in 20 minutes?

   (J) 420
   (K) 450
   (L) 500
   (M) 520

9. Jason wants to buy a stack of clothing that totals $400. The sales tax is 7%. If you add the tax to his total bill, what is Jason's grand total?

   (A) $407
   (B) $412
   (C) $417
   (D) $428

10. What is the Greatest Common Factor (GCF) of 60 and 72?

    (J) 12
    (K) 6
    (L) 3
    (M) 2

$$\frac{8}{10} - \frac{3}{5} + \frac{1}{2}$$

11. Evaluate the following expression above.

    (A) $\frac{6}{7}$
    (B) $\frac{10}{13}$
    (C) $\frac{7}{10}$
    (D) $\frac{6}{10}$

12. Paint Perfect Co. charges $55 every 100 square feet to paint a wall. If you have four walls with an area of 150 square feet each, how much will it cost to have the walls painted?

    (J) $165
    (K) $220
    (L) $275
    (M) $330

$$13 + 3 \times 2 + 16 \div (-4)$$

13. Evaluate the following expression above.

    (A) 9
    (B) 12
    (C) 15
    (D) 18

14. Binary fission is when a single cell splits in half to reproduce; for example, one cell becomes two, then the two cells become four. A scientist finds that a bacteria population count is 4 at 10:00am. If the bacteria population undergoes binary fission every 20 minutes, what is the best estimate of the population at 11:00am?

    (J) 16
    (K) 32
    (L) 36
    (M) 64

15. The price of a bag of popcorn is 25% of the price of a movie ticket. Jason paid $20 for 2 bags of popcorn and 2 movie tickets. What is the price of each bag of popcorn?

    (A) $2
    (B) $4
    (C) $6
    (D) $8

STOP

IAAT PRACTICE TEST 7

SECTION 2
Time — 10 minutes
15 Questions

Directions: Read the information given and choose the best answer for each question. Base your answer only on the information given. The time limit for each section is 10 minutes.

Directions: Use the following graph and information to answer questions 1 − 4.

There are 200 students in a school. 120 students are 5th graders and 80 students are 6th graders. Out of the 200 students, 100 students chose math as their favorite subject and the remaining 100 students chose English as their favorite subject.

	Math	English	Total
5th Graders	80		120
6th Graders			80
Total	100	100	200

1. How many 5th graders chose English as their favorite subject?

   (A) 20
   (B) 30
   (C) 35
   (D) 40

2. How many 6th graders chose math as their favorite subject?

   (J) 20
   (K) 30
   (L) 40
   (M) 80

3. How many 6th graders chose English as their favorite subject?

   (A) 20
   (B) 40
   (C) 60
   (D) 80

4. What percent of the 200 student population is 5th graders?

   (J) 25%
   (K) 40%
   (L) 60%
   (M) 65%

Directions: Use the following graph to answer questions 5 – 8.

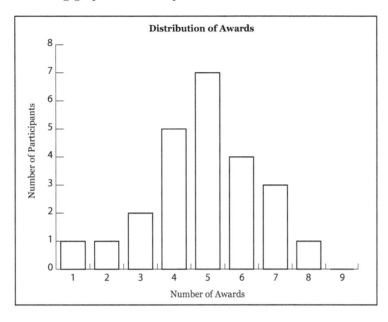

5. What is the mode number of awards received?

   (A) 4
   (B) 5
   (C) 7
   (D) 8

6. How many participants were there in all?

   (J) 7
   (K) 8
   (L) 24
   (M) 25

7. What percent of the participants received six awards?

   (A) $12\frac{1}{3}\%$
   (B) $16\frac{2}{3}\%$
   (C) $33\frac{1}{3}\%$
   (D) $66\frac{2}{3}\%$

8. How many participants received at most three awards?

   (J) 2
   (K) 4
   (L) 5
   (M) 9

Directions: Use the following table to answer questions 9 – 12. The number of computers sold at four stores from Week 1 through Week 5 is displayed.

	Store A	Store B	Store C	Store D
Week 1	18	12	25	35
Week 2	21	15	15	26
Week 3	8	10	19	31
Week 4	22	24	12	5
Week 5	22	15	28	18

9. Which store sold the most number of computers in week 4?

    (A) Store A
    (B) Store B
    (C) Store C
    (D) Store D

10. What is the total number of computers Store D sold during the 5-week period?

    (J) 105
    (K) 115
    (L) 130
    (M) 165

11. Store A had 12 returns after the report was given. What is the correct total number of computers sold by Store A during the 5-week period?

    (A) 91
    (B) 79
    (C) 72
    (D) 60

12. What is the mean number of computers sold by the four stores during Week 3.

    (J) 68
    (K) 17
    (L) 16
    (M) 14

13. Mr. Rhee is two-yards tall. Joshua is 4 feet 2 inches tall. Jason is 52 inches tall. Which of the following represents the order of height from shortest to tallest? (1 yard = 3 feet, and 1 foot = 12 inches)

    (A) Mr. Rhee, Joshua, Jason
    (B) Jason, Mr. Rhee, Joshua
    (C) Jason, Joshua, Mr. Rhee
    (D) Joshua, Jason, Mr. Rhee

14. Mr. Rhee ran ten miles on Monday. On Tuesday, he ran four miles less than he did on Monday. On Wednesday, he ran six miles more than he did on Tuesday. How many miles did Mr. Rhee run altogether?

    (J) 32
    (K) 28
    (L) 24
    (M) 20

15. Jason has three times the number of stickers as Sue. How many stickers must Jason give Sue so that they each have 40 stickers?

    (A) 20
    (B) 30
    (C) 35
    (D) 40

STOP

IAAT PRACTICE TEST 7

SECTION 3
Time — 10 minutes
15 Questions

Directions: Read the information given and choose the best answer for each question. Base your answer only on the information given. The time limit for each section is 10 minutes.

1. You currently do not have membership to a business center; however, a membership is required to use the internet. If the cost of membership is $2 and internet usage is charged at an hourly rate of $1.50, how long can you use the internet with $8?

   (A) 7
   (B) 6
   (C) 5
   (D) 4

2. Mr. Rhee has $500 in his bank account and spends $20 per day. If no additional income is added, which of the following equation represents the remaining amount of money, m, in his bank account after d days?

   (J) $m = 20d + 500$
   (K) $m = 20d - 500$
   (L) $m = 500 - 20d$
   (M) $d = 500 + 20m$

x	1	2	3	4	...	7
y	3	6	9	12	...	

3. Observe the numbers in the two columns below in order to determine which of the following belongs in the empty cell.

   (A) 21
   (B) 18
   (C) 15
   (D) 12

4. Which of the following table best represents the following verbal relationship? The number of guitars, y, sold is twelve less than three times the number of pianos, x, sold.

(J)
x	4	5	6	7
y	0	3	6	9

(K)
x	4	5	6	7
y	-8	-7	-6	-5

(L)
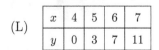

(M)

$$A + A + A = 36$$
$$B + B + A = 22$$

5. If A and B represent two different positive integers that satisfy both equation, which of the following is the value of B ?

   (A) 5
   (B) 4
   (C) 3
   (D) 2

6. Jason is an electronics sales associate. He makes a salary of $500 per week. In addition to his salary, he earns a $75 commission for every laptop he sells. Which of the following is an equation that represents the total amount of money, y, Jason earns per week if he sells x laptops?

   (J) $y = 500 + 75x$
   (K) $y = 500 - 75x$
   (L) $y = 75 + 500x$
   (M) $y = 75(x + 500)$

Input	Output
-1	-2
-3	-4
-5	-6

7. Above is a table that represents the x and y coordinates used to plot points on a coordinate system. Which of the Quadrants do all of the points lie within?

   (A) Quadrant I
   (B) Quadrant II
   (C) Quadrant III
   (D) Quadrant IV

8. Joshua has 3 fewer marbles than Mr. Rhee. Jason has twice as many marbles as Joshua. If Mr. Rhee has m marbles, which of the following is an expression that represents the number of marbles that Jason has?

   (J) $2m + 3$
   (K) $2m + 6$
   (L) $2m - 3$
   (M) $2m - 6$

9. The average of two numbers is 10. If a third number is added to the set, the average of the three numbers is 12. What number was added to the set?

   (A) 12
   (B) 13
   (C) 15
   (D) 16

10. A pizza delivery company charges $7.00 per pizza and a flat delivery charge of $2.50. Which of the following is an equation that represents the total amount of money, T, spent when ordering x pizzas for delivery?

   (J) $T = 7x - 2.5$

   (K) $T = 7x + 2.5$

   (L) $T = 2.5 - 7x$

   (M) $T = \frac{7}{2.5}x$

11. Which of the following line does NOT represent a function?

   (A)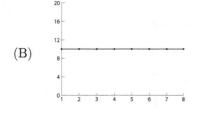

   (B)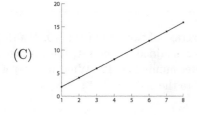

   (C)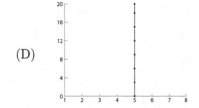

   (D)

$$y = 2x + 3$$

12. Which ordered pair is a solution to the equation above?

   (J) $(3, 3)$

   (K) $(3, -3)$

   (L) $(-3, 3)$

   (M) $(-3, -3)$

y is 3 less than one half the value of x

13. Write the verbal relationship as an equation.

   (A) $y = \frac{x}{2} - 3$

   (B) $y = \frac{1}{2}x + 3$

   (C) $y = 3x - \frac{1}{2}$

   (D) $y = 3 - \frac{1}{2}x$

14. Jason has $9 to spend at a carnival. Each game requires $0.75 per play and a fountain soda costs $2.50. Which of the following expression represents the amount of money Jason has remaining after purchasing y drinks and playing x games?

(J) $9 + 2.5y + 0.75x$

(K) $2.5y + 0.75x - 9$

(L) $9 - 2.5y - 0.75x$

(M) $9 - 2.5x - 0.75y$

15. The area of a circle is defined as πr^2; where r represents the radius. If the radius of the circle A and B are 2 and 3 respectively, how much larger is the area of circle B than that of circle A?

(A) 5π

(B) 6π

(C) 7π

(D) 8π

STOP

IAAT PRACTICE TEST 7

SECTION 4
Time — 10 minutes
15 Questions

Directions: Read the information given and choose the best answer for each question. Base your answer only on the information given. The time limit for each section is 10 minutes.

1. If Sue walks $\frac{1}{2}$ mile per hour, which expression would represent the distance she walks in x hours?

    (A) $\frac{1}{2}x$

    (B) $\frac{1}{2}(60)x$

    (C) $x + \frac{1}{2}$

    (D) $60x + \frac{1}{2}$

2. If a jug of coffee contains 8 servings and each serving is 150 calories, which of the following is the equation that represents the total number of calories, c, in the jug?

    (J) $c = \frac{8}{150}$

    (K) $c = 8 + 150$

    (L) $c = \frac{150}{8}$

    (M) $c = 8 \times 150$

3. Solve the inequality: $\frac{x}{4} > 6$

    (A) $x > \frac{6}{4}$

    (B) $x > 2$

    (C) $x < \frac{3}{2}$

    (D) $x > 24$

4. If x is a positive integer, which of the following cannot be a value of x^2?

    (J) 2

    (K) 4

    (L) 16

    (M) 81

5. If $x = 4$, what is the value of $\frac{x^3}{x}$?

    (A) 4

    (B) 16

    (C) 64

    (D) 256

6. Let $c^2 = a^2 + b^2$, where a, b, and c are positive integers. If $a = 6$ and $b = 8$, what is the positive integer value of c?

    (J) 100

    (K) 14

    (L) 10

    (M) 9

The quotient of 35 and a number, r, is five

7. Solve for r in the verbal phrase stated above.

   (A) 7
   (B) 30
   (C) 35
   (D) 175

$$5y - 2 = -4y - x$$

8. Rewrite the equation so that y is a function of x.

   (J) $y = \frac{2-x}{9}$
   (K) $y = -x + 2$
   (L) $y = x - 2$
   (M) $x = 9y + 2$

---

9. Jason can finish 14 math problems in 10 minutes. At the same rate, which of the following proportion can be used to find p, the number of problems Jason can complete in 45 minutes?

   (A) $\frac{10}{45} = \frac{p}{14}$
   (B) $\frac{10}{14} = \frac{45}{p}$
   (C) $\frac{p}{14} = \frac{10}{45}$
   (D) $\frac{p}{45} = \frac{10}{14}$

---

10. What is the value of $x^3 + x$ when $x = 3$?

    (J) 12
    (K) 27
    (L) 30
    (M) 81

$$\frac{x}{2} = \frac{y}{3} = \frac{z}{4} = 2$$

11. What is the value of $x + y + z$?

    (A) 24
    (B) 18
    (C) 12
    (D) 6

---

12. Solve for x: $5x - 15 = 85$

    (J) 1
    (K) 7
    (L) 14
    (M) 20

---

13. If $x = -1.5$ and $y = 2$, what is $5y - 4x$?

    (A) 4
    (B) 6
    (C) 10
    (D) 16

$\{(-1, 0), (0, 1), (1, 2), (2, 3)\}$

14. A set of four ordered pairs shown above represents a function. Which of the following set of numbers represents the domain of the function?

    (J) $\{-1, 0, 1, 2\}$
    (K) $\{0, 1, 2, 3\}$
    (L) $\{-1, 1, 2\}$
    (M) $\{1, 2, 3\}$

15. If $|x| = 4$, which of the following could be the value of x?

    (A) -2
    (B) -3
    (C) -4
    (D) -5

STOP

SOLOMON ACADEMY   Distribution or replication of any   TEST 7 SECTION 1
                        part of this page is prohibited.

Answers and Solutions
IAAT Practice Test 7 Section 1

Answers

1. D	2. J	3. C	4. L	5. C
6. J	7. B	8. L	9. D	10. J
11. C	12. M	13. C	14. K	15. A

Solutions

1. (D)

   In order to divide fractions, multiply the first fraction by the reciprocal, or the multiplicative inverse, of the second fraction. The reciprocal of $\frac{1}{3}$ is $\frac{3}{1}$. Furthermore, convert the mixed number $4\frac{2}{3}$ into an improper fraction, $4\frac{2}{3} = \frac{4 \times 3 + 2}{3} = \frac{14}{3}$. Thus,

   $$4\frac{2}{3} \div \frac{1}{3} = \frac{14}{3} \times \frac{3}{1} = 14$$

   Therefore, (D) is the correct answer.

2. (J)

   Since $0.5 = \frac{1}{2}$, $15 \div 0.5$ can be rewritten as $15 \div \frac{1}{2}$. In order to divide fractions, multiply the first fraction by the reciprocal, or the multiplicative inverse, of the second fraction.

   $$15 \div 0.5 = 15 \div \frac{1}{2} = 15 \times 2 = 30$$

   Therefore, (J) is the correct answer.

3. (C)

   There are 60 seconds in one minute. Thus, there are $3 \times 60 = 180$ seconds in 3 minutes. Since it takes 9 seconds for the scanner to scan 1 page of a document, the scanner will scan $\frac{180}{9} = 20$ pages in 3 minutes.

4. (L)

Scientific notation must be written in the form: $c \times 10^n$, where $1 \leq c < 10$ and n is an integer. In general, positive n values give larger values than 10 and negative n values give smaller values than 1.

$$\begin{aligned}
3 \times 10^3 \times 5 \times 10^2 &= (3 \times 5) \times (10^3 \times 10^2) \\
&= 15 \times 10^5 \qquad \text{(Since } 15 = 1.5 \times 10\text{)} \\
&= 1.5 \times 10 \times 10^5 \qquad \text{(Since } 10^1 \times 10^5 = 10^6\text{)} \\
&= 1.5 \times 10^6
\end{aligned}$$

Therefore, (L) is the correct answer.

5. (C)

In order to find the total amount Joshua spent during the four months, add up all of the values given. Therefore, $\$23 + \$25 + \$55 + \$57 = \$160$.

6. (J)

In order to add rational numbers in the form of a fraction with uncommon denominators, it is necessary to find the least common denominator (LCD). The least common multiple of 21 and 84 is 84. Multiply both the numerator and denominator of the first fraction by 4 in order to make both fractions have a common denominator.

$$\frac{6}{21} - \frac{23}{84} = \frac{24}{84} - \frac{23}{84} = \frac{1}{84}$$

Therefore, (J) is the correct answer.

7. (B)

The original cost of the computer is $1000 and Jason purchased it for $750. This means that Jason paid $\frac{750}{1000} = 0.75$ or 75% the original cost of the computer. Therefore, the sale percentage off the computer is $100\% - 75\% = 25\%$.

8. (L)

Joshua can walk 5 blocks in 4 minutes. At the same rate, by using proportions, Joshua walks 25 blocks in 20 minutes. Since each block is 20 feet long, the total distance that Joshua walks in 20 minutes is $25 \times 20 = 500$ feet.

9. (D)

Jason wants to buy a stack of clothing that totals $400 before tax. The sales tax is 7%. This means that the total bill will be 107% of $400. Out of 107%, 100% represents the original cost of clothes and 7% represents the sales tax. Since $107\% = 1.07$, the total amount of the bill is $400 \times 1.07 = 428$.

10. (J)

It is possible to solve for the Greatest Common Factor (GCF) by listing out the factors for each number or by prime factorization.

Factors of 60: 1, 2, 3, 4, 5, 6, 10, **12**, 15, 20, 30, and 60
Factors of 72: 1, 2, 3, 4, 6, 8, 9, **12**, 18, 24, 36, and 72
Since 12 is the largest factor of both 60 and 72, 12 is the GCF of 60 and 72.

The prime factorization of 60: $\mathbf{2 \times 2 \times 3} \times 5$
The prime factorization of 72: $\mathbf{2 \times 2} \times 2 \times \mathbf{3} \times 3$
Since there are at least two 2's and one 3 in both the prime factorization of 60 and 72, the GCF is $2 \times 2 \times 3 = 12$.

Therefore, (J) is the correct answer.

11. (C)

In order to add or subtract fractions, it is necessary to have a common denominator. The least common multiple of 10, 5, and 2 is 10. Thus, the least common denominator of $\frac{8}{10}$, $\frac{3}{5}$, and $\frac{1}{2}$ is 10. After converting each denominator to 10, evaluate the expression.

$$\frac{8}{10} - \frac{3}{5} + \frac{1}{2} = \frac{8}{10} - \frac{6}{10} + \frac{5}{10} = \frac{8 - 6 + 5}{10} = \frac{7}{10}$$

Therefore, (C) is the correct answer.

12. (M)

Paint Perfect Company charges $55 every 100 square feet of work. The four walls, each with an area of 150 square feet, has a total area of $150 \times 4 = 600$ square feet. The amount needed to be painted is $\frac{600}{100} = 6$ times the charged rate of work. Therefore, the total cost of painting these four walls is $\$55 \times 6 = \$330$.

13. (C)

When evaluating expressions, remember the order of operations: PEMDAS. Solve all expressions in a parenthesis first, then exponents, then multiplication and division from left to right, and finally addition and subtraction from left to right.

$$13 + 3 \times 2 + 16 \div (-4) = 13 + 6 - 4 = 15$$

Therefore, (C) is the correct answer.

14. (K)

A scientist finds that a bacteria population count is 4 at 10:00am and doubles in size every 20 minutes. In order to find the bacteria population count at 11:00am, it is necessary to find out the number of 20 minute intervals between 10:00am and 11:10am. There are 3 intervals between 10:00am and 11:10am: 10:00-10:20, 10:20-10:40, and 10:40-11:00. Since the bacteria count doubles every 20 minutes, the bacteria count is $2 \times 2 \times 2$ or $2^3 = 8$ times the original value after an hour. Therefore, the best estimated value of the bacteria population at 11:00am is $4 \times 8 = 32$.

15. (A)

The price of a bag of popcorn is 25% of the price of a movie ticket. This means that the price of 4 bags of popcorn is equal to the price of one movie ticket. The price of 2 movie tickets is equal to the price of $2 \times 4 = 8$ bags of popcorn. Furthermore, the price of 2 bags of popcorn and 2 movie tickets is equal to $2 + 8 = 10$ bags of popcorn. Since Jason paid $20, the price of each bag of popcorn is $\frac{\$20}{10} = \$2$.

SOLOMON ACADEMY　　Distribution or replication of any part of this page is prohibited.　　TEST 7 SECTION 2

Answers and Solutions
IAAT Practice Test 7 Section 2

Answers

1. D	2. J	3. C	4. L	5. B
6. L	7. B	8. K	9. B	10. K
11. B	12. K	13. D	14. K	15. A

Solutions

1. (D)

   The following table has been filled in to help solve questions 1-4.

	Math	English	Total
5th Graders	80	**40**	120
6th Graders	**20**	60	80
Total	100	100	200

   Since there is a total of 120 5th graders, of which 80 favored math, there are $120 - 80 = 40$ 5th graders who favored English.

2. (J)

   There are 100 people in the school that favored math as their favorite subject, of which 80 were 5th graders. Therefore, the number of 6th graders who chose math as their favorite subject is $100 - 80 = 20$.

3. (C)

   After filling out the table, it is possible to determine how many 6th graders chose English as their favorite subject. 20 of the 80 6th graders chose math as their favorite subject. Therefore, the remaining $80 - 20 = 60$ 6th graders chose English.

4. (L)

   A percent, 1 %, means 1 out of 100 or $\frac{1}{100}$. Out of the 200 students, 120 students are 5th graders. In order to find the percentage, set up a fraction of the number of 5th graders to the total number of students. Afterwards, set the denominator to 100 and the numerator represents the percentage.

   $$\text{Percent} = \frac{120}{200} = \frac{60}{100} = 60\%$$

   Therefore, 60% of the 200 students are the 5th graders.

5. (B)

The mode of a data set represents the element which appears the most. The y-axis is a representation of how many participants received a specific number of awards. Since there were 7 participants who received 5 awards, the mode number of awards received is 5. Furthermore, this can be determined visually as 5 awards has the highest peak. Therefore, (B) is the correct answer.

6. (L)

The total number of participants can be determined by observing the y-value for every x-value because the y-axis represents how many participants received a specific number of awards. Therefore, the total number of participants is $1 + 1 + 2 + 5 + 7 + 4 + 3 + 1 = 24$.

7. (B)

Percent can be determined by the part over the whole. In order to convert a decimal into a percentage, multiply by 100 or move the decimal point two places to the right. Out of 24 participants, four people received 6 awards. Therefore, the percent of participants who received 6 awards is $\frac{4}{24} = \frac{1}{6} = 16\frac{2}{3}\%$.

8. (K)

At most 3 awards means that the participants received 3 or less awards. Since 2 people received 3 awards, 1 person received 2 awards, and 1 person received 1 award, there are 4 participants who had at most 3 awards.

9. (B)

During week 4, store A sold 22 computers, store B sold 24 computers, store C sold 12 computers, and store D sold 5 computers. Therefore, the store that sold the most number of computers in week 4 is store B.

10. (K)

The total number of computers store D sold during the 5-week period is $35 + 26 + 31 + 5 + 18 = 115$ computers.

11. (B)

During the 5-week period, store A sold $18 + 21 + 8 + 22 + 22 = 91$ computers of which 12 were returned. Therefore, the total number of computers sold by store A after 12 returns is $91 - 12 = 79$ computers.

12. (K)

In order to find the mean, or average, divide the total sum of elements by the number of elements. In other words, divide the total number of computers collectively sold by all four stores during week 3 by 4.

$$\text{Average} = \frac{8 + 10 + 19 + 31}{4} = \frac{68}{4} = 17$$

Therefore, the mean number of computers sold by the four stores during Week 3 is 17.

13. (D)

Convert all of the information given into the same units for easy comparison. All of the units were converted to inches for this example. Mr. Rhee is two-yards tall. Since one yard is equivalent to 3 feet and one feet is equivalent to 12 inches, Mr. Rhee is $2 \times 3 \times 12 = 60$ inches tall. Joshua is 4 feet 2 inches tall which means that he is $4 \times 12 + 2 = 48 + 2 = 50$ inches tall. Jason is 52 inches tall which was given in the question. Thus, the order of height from shortest to tallest is Joshua, Jason, Mr. Rhee. Therefore, (D) is the correct answer.

14. (K)

Mr. Rhee ran ten miles on Monday. On Tuesday, he ran four miles less than he did on Monday. This means that Mr. Rhee ran $10 - 4 = 6$ miles on Tuesday. On Wednesday, he ran six miles more than he did on Tuesday. This means that Mr. Rhee ran $6 + 6 = 12$ miles on Wednesday. Therefore, the total distance Mr. Rhee ran during these 3 days is $10 + 6 + 12 = 28$ miles.

15. (A)

If they each will have 40 stickers at the end, it means that there are $40 \times 2 = 80$ stickers in total. Since Jason has three times the number of stickers as Sue, it means that the ratio of Jason's to Sue's number of stickers is $3 : 1$. Thus, let $3x$ and x be the number of stickers Jason and Sue have respectively.

$$3x + x = 80 \qquad \text{(Combine like terms)}$$
$$4x = 80 \qquad \text{(Divide each side by 4)}$$
$$x = 20$$

This means that Sue initially had 20 stickers and Jason initially had $20 \times 3 = 60$ stickers. Thus, Jason needs to give Sue 20 stickers so that they each have 40 stickers. Therefore, (A) is the correct answer.

SOLOMON ACADEMY TEST 7 SECTION 3

Answers and Solutions
IAAT Practice Test 7 Section 3

Answers

1. D	2. L	3. A	4. J	5. A
6. J	7. C	8. M	9. D	10. K
11. D	12. M	13. A	14. L	15. A

Solutions

1. (D)

    The cost of a membership is $2. Since you only have $8, after the cost of membership you will have $8 − $2 = $6. Internet usage is charged at an hourly rate of $1.50. Therefore, the total number of hours you can use the internet with the remaining $6 is $\frac{6}{1.5} = 4$ hours.

2. (L)

    Mr. Rhee has $500 in his bank account and spends $20 per day. In other words, the initially amount of money that Mr. Rhee has is $500 and that the amount decreases by $20 per day. The amount decreases by $20 per day for d days can be expressed as $-20d$. Therefore, the best equation which represents this statement is $m = 500 - 20d$.

3. (A)

    The table represent a function that can be defined by the equation $y = 3x$. Therefore, when $x = 7$, the value of y is $3(7) = 21$.

4. (J)

x	$y = 3x - 12$
4	$y = 3(4) - 12 = 0$
5	$y = 3(5) - 12 = 3$
6	$y = 3(6) - 12 = 6$
7	$y = 3(7) - 12 = 9$

    The number of guitars, y, sold is twelve less than three times the number of pianos, x, sold can be represented by the equation $y = 3x - 12$. Make sure when observing ordered pairs that all ordered pairs satisfy the equation. As shown in the table above, all the ordered pairs in answer choice (J) satisfy the equation and is the correct answer.

5. (A)

$A + A + A = 3A = 36$. Thus, $A = \frac{36}{3} = 12$. Substitute 12 for A in the second equation and solve for B. When solving for a variable, use the reverse order of operations, SADMEP, and inverse operations.

$$\begin{aligned} B + B + A &= 22 & \text{(Substitute 12 for } A \text{ and combine like terms)} \\ 2B + 12 &= 22 & \text{(Subtract 12 from each side)} \\ 2B &= 10 & \text{(Divide each side by 2)} \\ B &= 5 \end{aligned}$$

Therefore, the value of B is 5.

6. (J)

Jason makes a fixed salary of $500 per week plus an additional $75 per laptop he sells. If Jason sells x number of laptops, he earns sales commission, which can be expressed as $75x$. Therefore, the equation that represents the total amount of money, y, Jason earns per week if he sells x number of laptops is $y = 500 + 75x$.

7. (C)

The first quadrant has ordered pairs with positive x and y values. The second quadrant has ordered pairs with negative x values but positive y values. The third quadrant has ordered pairs with negative x and y values. Lastly, the fourth quadrant has ordered pairs with positive x values but negative y values. Since all the ordered pairs represented in the table have negative values, they are the ordered pairs in the third quadrant. Therefore, (C) is the correct answer.

8. (M)

Mr. Rhee has m number of marbles. Since Joshua has 3 fewer marbles than Mr. Rhee, Joshua has $m - 3$ number of marbles. Jason has twice as many marbles as Joshua. Thus, Jason has $2(m - 3)$ marbles. In order to expand the expression $2(m - 3)$, use the distributive property: $a(b-c) = ab - ac$. Thus, the number of marbles that Jason has is $2(m-3) = 2m - 2(3) = 2m - 6$. Therefore, (M) is the correct answer.

9. (D)

When dealing with average, it may be necessary to find the sum to help you find the answer. The sum of a data set is equivalent to the product of the average and the number of elements. For example, since the average of two numbers is 10, the sum of those two numbers is $10 \times 2 = 20$. If a third number is added to the set, the average of the three numbers is 12. Since the average of the three numbers is 12, the sum of those three numbers is $12 \times 3 = 36$. In order to find the number which was added to the set, subtract the sum of the three numbers by the sum of the two numbers. Therefore, the number which was added to the set is $36 - 20 = 16$.

10. (K)

Since it costs $7 per pizza, it costs $7x$ to buy x number of pizzas. There is a flat delivery charge of $2.50 which is unchanging regardless of how many pizzas you order. Thus, the equation that best represents the total amount of money, T, spent when ordering x pizzas for delivery is $T = 7x + 2.5$.

11. (D)

A function relates an input to an output. Thus, an input x cannot have more than one value for an output y. When a line graph is given, it is possible to determine whether or not it is a function by the vertical line test. If a vertical line can be drawn at any location and crosses through multiple points, it fails the vertical line test and is NOT a function. This is evident in answer choice (D), as the line depicted is a vertical line. Therefore, answer choice (D) is NOT a function.

12. (M)

Plug in the different ordered pairs into the given equation, $y = 2x + 3$, to see which makes the equation true. An ordered pair describes the location of a point and is written in the form (x, y) where the first number represents the x-coordinate and the second number represents the y-coordinate.

$$y = 2x + 3 \quad \text{(Substitute } (-3, -3) \text{ into the equation)}$$
$$-3 = 2(-3) + 3$$
$$-3 = -6 + 3$$
$$-3 = -3$$

Since the ordered pair $(-3, -3)$ satisfies the equation $y = 2x + 3$, it is a solution to the equation. Therefore, (M) is the correct answer.

13. (A)

Verbal Phrase	Expression
One half the value of x	$\frac{x}{2}$
3 less than one half the value of x	$\frac{x}{2} - 3$
y is 3 less than one half the value of x	$y = \frac{x}{2} - 3$

Therefore, (A) is the correct answer.

14. (L)

Jason has $9 to spend at a carnival. Each fountain soda costs $2.50 and Jason wants to purchase y number of sodas. This can be expressed as $2.5y$. Furthermore, each game requires $0.75 per play. If Jason plays x games, the total cost for the carnival games would be $0.75x$. Thus, the amount of money Jason has remaining after purchasing y drinks and playing x games is $9 - 2.5y - 0.75x$. Therefore, (L) is the correct answer.

15. (A)

The area of a circle is defined as πr^2, where r is the radius. There are two circles A and B with the radius of 2 and 3, respectively. The area of circle B is $\pi(3)^2 = 9\pi$. The area of circle A is $\pi(2)^2 = 4\pi$. Thus, the area of circle B is $9\pi - 4\pi = 5\pi$ larger than the area of circle A. Therefore, (A) is the correct answer.

SOLOMON ACADEMY — TEST 7 SECTION 4

Answers and Solutions
IAAT Practice Test 7 Section 4

Answers

1. A	2. M	3. D	4. J	5. B
6. L	7. A	8. J	9. B	10. L
11. B	12. M	13. D	14. J	15. C

Solutions

1. (A)

   If Sue walks $\frac{1}{2}$ mile in one hour, she walks $\frac{1}{2}x$ miles in x hour. Therefore, (A) is the correct answer.

2. (M)

   A jug of coffee contains 8 servings and each serving is 150 calories. To determine the total number of calories in the jug, use multiplication. Therefore, the equation that represents the total number of calories, c, in the jug is $c = 8 \times 150$.

3. (D)

   In order to solve for the variable x, multiply each side of the inequality by 4.

   $$\frac{x}{4} > 6 \qquad \text{(Multiply each side by 4)}$$

   $$x > 24$$

   Therefore, (D) is the correct answer.

4. (J)

   Since $2^2 = 4$, $4^2 = 16$, and $9^2 = 81$, 4, 16, and 81 can be the value of x^2. However, 2 cannot be the value of x^2 since x is a positive integer. Therefore, (J) is the correct answer.

5. (B)

   Recall the property of exponent: $\frac{a^m}{a^n} = a^{m-n}$. Thus, $\frac{x^3}{x} = \frac{x^3}{x^1} = x^{3-1} = x^2$. Since $x = 4$, $x^2 = 4^2 = 16$. Therefore, the value of $\frac{x^3}{x}$ when $x = 4$ is 16.

6. (L)

$$c^2 = a^2 + b^2$$ (Substitute 6 for a and 8 for b)
$$c^2 = 6^2 + 8^2$$ (Divide both sides by 3)
$$c^2 = 36 + 64$$
$$c^2 = 100$$ (Since c is a positive integer)
$$c = 10$$

Thus, the positive integer value of c is 10.

7. (A)

The quotient of 35 and a number, r, is five can be represented by the equation $\frac{35}{r} = 5$. When solving for the variable r, use the reverse order of operations, SADMEP, and inverse operations.

$$\frac{35}{r} = 5$$ (Multiply each side by r)
$$35 = 5r$$ (Divide each side by 5)
$$r = 7$$

Therefore, the value of r is 7.

8. (J)

Rewrite the equation so that y is a function of x, which means to solve for y.

$$5y - 2 = -4y - x$$ (Add $4y$ to each side)
$$9y - 2 = -x$$ (Add 2 to each side)
$$9y = 2 - x$$ (Divide each side by 9)
$$y = \frac{2-x}{9}$$

Therefore, (J) is the correct answer.

9. (B)

The question states that Jason can finish 14 math problems in 10 minutes. At the same rate means that it is possible to use proportions to determine how many math problems Jason can finish in 45 minutes. Proportions state that two ratios, usually expressed as fractions, are the same.

$$\frac{\text{Minutes}}{\text{Questions Finished}} : \frac{10}{14} = \frac{45}{p}$$

The proportion in answer choice (B) is correct because it represents a ratio of time to the number of questions finished. Therefore, (B) is the correct answer.

10. (L)

When $x = 3$, the value of x^3 is $3^3 = 3 \times 3 \times 3 = 27$. Therefore, the value of $x^3 + x$ is $27 + 3 = 30$.

11. (B)

    If $\frac{x}{2} = 2$, then $x = 4$. If $\frac{y}{3} = 2$, then $y = 6$. If $\frac{z}{4} = 2$, then $z = 8$. Therefore, the value of $x + y + z$ is $4 + 6 + 8 = 18$.

12. (M)

    In order to solve for x, use the reverse order of operations, SADMEP, and inverse operations.

    $$5x - 15 = 85 \qquad \text{(Add 15 to each side)}$$
    $$5x = 100 \qquad \text{(Divide each side by 5)}$$
    $$x = 20$$

    Therefore, the solution to $5x - 15 = 85$ is $x = 20$.

13. (D)

    Substitute -1.5 for x and 2 for y and evaluate the expression.

    $$5y - 4x = 5(2) - 4(-1.5) = 10 + 6 = 16$$

    Therefore, (D) is the correct answer.

14. (J)

    A set of four ordered pairs, $\{(-1, 0), (0, 1), (1, 2), (2, 3)\}$, represents a function because x-values are not repeated. Since the domain of a function is a set of x-values, the domain of the function is $\{-1, 0, 1, 2\}$. Therefore, (J) is the correct answer.

15. (C)

    Absolute value represents the distance a number is from zero. Thus, $|-4| = 4$ because -4 is 4 units from zero. Therefore, (C) is the correct answer.

CPSIA information can be obtained
at www.ICGtesting.com
Printed in the USA
LVHW061238280319
611858LV00023B/21/P